Lukas Wengeler

Coating and drying processes for functional films in polymer solar cells

From laboratory to pilot scale

Coating and drying processes for functional films in polymer solar cells

From laboratory to pilot scale

by
Lukas Wengeler

Dissertation, Karlsruher Institut für Technologie (KIT)
Fakultät für Chemieingenieurwesen und Verfahrenstechnik
Tag der mündlichen Prüfung: 14. März 2014
Referent: Prof. Dr.-Ing. Dr. h. c. Wilhelm Schabel
Korreferent: Prof. Dr.-Ing. habil. Hermann Nirschl

Impressum

 Scientific Publishing

Karlsruher Institut für Technologie (KIT)
KIT Scientific Publishing
Straße am Forum 2
D-76131 Karlsruhe

KIT Scientific Publishing is a registered trademark of Karlsruhe
Institute of Technology. Reprint using the book cover is not allowed.

www.ksp.kit.edu

Print on Demand 2014

ISBN 978-3-7315-0201-2
DOI: 10.5445/KSP/1000039966

COATING AND DRYING PROCESSES FOR FUNCTIONAL FILMS IN POLYMER SOLAR CELLS

- FROM LABORATORY TO PILOT SCALE

Zur Erlangung des akademischen Grades eines

DOKTORS DER INGENIEURWISSENSCHAFTEN (Dr.-Ing.)

Von der Fakultät für Chemieingenieurwesen und Verfahrenstechnik des

Karlsruher Institut für Technologie (KIT)

genehmigte

Dissertation

von Dipl.-Ing. Lukas Wengeler

geboren in Sandton, Südafrika.

Referent:	Prof. Dr.-Ing. Dr. h. c. Wilhelm Schabel
Korreferent:	Prof. Dr.-Ing. habil. Hermann Nirschl
Tag der mündlichen Prüfung:	14.03.2014

Preface

The research for this thesis was conducted from 2009 to 2013 during my work as research assistant at the Institute of Thermal Process Engineering (TVT) at the Karlsruher Institute of Technology (KIT).

First of all, I want to thank Prof. Dr.-Ing. Dr. h. c. Wilhelm Schabel for the supervision of this thesis and Dr.-Ing. Philip Scharfer for his guidance. I especially appreciate their confidence and enjoyed the opportunity to work independently, implement own ideas, and present my work at several meetings and conferences.

I would like to thank Prof. Dr.-Ing. Matthias Kind and Prof. Dr.-Ing. Thomas Wetzel for the support of the institute. Special thanks go to the technical staff: Michael Wachter, Markus Keller, and Steffen Haury. The manufacture of specialized coating tools and the customization of the experimental set-up was invaluable for the success of this thesis. Gisela Schimana and Margit Morvay at the TVT secretary deserve credit for helping me with administrative work. I would like to thank all colleagues at the institute, not only for their help and support, but also for the social and friendly atmosphere which made working in Karlsruhe such a pleasant experience.

For their dedicated work during student research projects and research jobs I thank Fabian Brunner, Marcel Schmitt, Stefan Jaiser, Konrad Döring, Peter Darmstädter, Martin Grunert, Christian Kroner, Christian Hotz, Thomas Wenz, Jana Kumberg, Ralf Diehm, and Harald Wiegand. Working together with these students was a source of motivation for me.

This work was conducted in close collaboration with other PhD students at the Institute. I would like to thank Dipl.-Ing. Katharina Peters for her work on experimental equipment and procedures, the start-up of the technical laboratory coater, and many scientific discussions. I am grateful to Dipl.-Ing. Marcel Schmitt, Dipl.-Ing. Philipp Cavadini, Dipl.-Ing. Felix Buss, Dipl.-Ing. Michael Baunach, and Dipl.-Ing. Stefan Jaiser for their cooperation. Their contributions are referenced throughout this work.

Prof. Dr. Uli Lemmer, Dr.-Ing. Dipl.-Phys. Alexander Colsmann, Manuel Reinhardt, Felix Nickel and the organic photovoltaics group at the Light Technology Institute (LTI) deserve thanks for their theoretical and practical help with solar cell devices. Dr. Stefan Wahlheim, Dr. Norman Mechau, and Jörg Pfeiffer provided access to profiler, evaporator, atomic force microscope and spin coater at the Institute of Nano Technology (INT) for which I am grateful. I thank Dr. Hendrik Hölscher for letting me use the atomic force microscope at the Karlsruhe Nano Micro Facility (KNMF).

I would like to thank Dr. Karen Köhler, Dr. Alexandra Große-Böwing, Dr. Phenwisa Niyamakom, Sasha Plug, Dr. Werner Hoheisel, and Dr. Frank Rauscher for providing materials, characterizing solar cells, and in general for the collaboration of Bayer Technology Services GmbH. Thanks go to Dr. Henry Wilson of Merck Chemicals Ltd. and Jamie Leganes at Innovation Lab GmbH for providing materials and the characterization of solar cells.

In particular, I want to thank my parents for their support throughout my academic career. My family and my wife provided the mental support and distraction that was necessary to finish this thesis with a smile on my face.

Finally, I acknowledge the financial support of the Federal Ministry for Education and Research (BMBF) within the research projects "Nanoteilchen basierte Polymersolarzellen" and "Print OLED 2". The research project "Druckbare Elektroden für effiziente organische Solarzellen" was financially supported by the Baden-Württemberg Stiftung.

Abstract

The supply of clean, renewable energy is one of the most pressing environmental challenges. Photovoltaic technologies are promising due to the abundant availability of solar energy and its direct conversion to electric energy. Whereas conventional solar cell technologies struggle to achieve higher efficiencies and lower prices, polymer based solar cells have continuously improved their performance throughout the last years and can be produced by a continuous coating process on low cost substrate. However, films with thickness of few nanometers have to be prepared over lateral dimensions of several centimeters with high demand on film quality. In addition, a complex interaction between process parameters, film properties, and device efficiency requires a thorough optimization of the production process. An experimental set-up needs to allow for a precise adjustment of process parameters and environmental conditions but provide compatibility to a large scale coating and drying process as well.

In this work an intermediate scale set-up is presented that enables slot-die coating and impingement drying under inert or clean room conditions. The batch-wise processing allows for the use of flexible as well as rigid substrates. Key features are a round nozzle drier with local extraction of spent air and a slot-die coating system with high precision gap adjustment and minimized hold-up. The same slot-die system can be used in a pilot-scale roll-to-roll line, facilitating the scale-up. The pilot coating line was developed with the machine manufacturer and set-up in the coating and printing laboratory at KIT campus north. It features a modular design for pretreatment, coating, and drying, which can be easily adapted to specific requirements. A gravure coating system was implemented to coat either 5 gravure test patterns at the same time; or a single pattern with as little as 3 ml coating ink. For the design of coating tools and the discussion of results, fluid dynamic properties have to be known.

Here, experimental data for the viscosity of common material systems are presented and fitted with empirical models as a function of composition, temperature and shear-rate. Surface tension, determined mainly by solvent composition, is discussed for typical solvent compositions. For water based dispersions of conductive polymer, the high surface tension of the ink can cause de-wetting of the film. The modification of surface tension by the addition of surfactant and the activation of surface energy by plasma treatment are described quantitatively.

To evaluate the potential of different coating methods, films were produced by spin, knife, spray, and slot-die coating, employing the same ink and as far as possible identical process conditions. The stability of slot-die coating was described by a dimensionless coating window for inks with variable viscosity and surface tension. Dry films were characterized using atomic force microscopy and UV-Vis spectroscopy. Both, surface topography and absorption spectra of films coated from dispersions, revealed no significant impact of coating method. In contrast, the photoactive layers coated from a polymer fullerene solution showed a more pronounced surface topography and an increase in light absorption for the spin coated samples. This result can be explained by an alignment of polymers due to shear forces in the coating flow. This effect may significantly contribute to reduced efficiency of cells produced in pilot scale processes compared to spin coated cells. To test this hypothesis, different shear-rates were introduced by slot-die coating at varying coating speed and slot width. The results confirmed that higher shear rates result in an increased light absorption of the photoactive layer, most pronounced for low boiling point solvents. It is thus possible to intentionally increase the shear rate in a pilot scale process to improve the optical properties of the photoactive films.

The impact of drying rate on the opto-electric properties of photoactive layers has been studied in detail under laboratory conditions. In this work, drying process parameters were investigated under technical conditions where transfer coefficients are significantly higher and isothermal conditions do not apply. Air flow rate and drying temperature had no or only a moderate impact on light absorption. The change of solvent composition to a higher fraction of high-boiling-point solvent resulted in an increase in light absorption but in a decrease in film stability at the same time.

Solar cells with hybrid photoactive layer were prepared by spray, knife, and slot-die coating, demonstrating the scalability of the technology. Slot-die coated polymer fullerene solar cells were produced on glass with the batch coater and on PET with the pilot coating line. Multilayer stacks with fully liquid processed electrode and adaptation layers were produced, showing promising film formation and the potential for low cost solar cells production.

Kurzdarstellung

Die Erschließung erneuerbarer Energiequellen ist eine der wichtigsten ökologischen Herausforderungen. Photovoltaische Technologien sind aufgrund der großen Verfügbarkeit solarer Energie und deren direkten Umwandlung in elektrische Energie vielversprechend. Wo eine weitere Steigerung von Wirkungsgrad oder Reduktion der Kosten bei konventionellen Solarzellen schwierig ist, verzeichnen Polymersolarzellen in den letzten Jahren stetig wachsende Leistung und besitzen das Potential mittels kontinuierlicher Beschichtungsverfahren auf kostengünstigen Substraten hergestellt zu werden. Filme mit Schichtdicken von wenigen Nanometern müssen dazu über laterale Dimensionen von mehreren Zentimetern bei hohen Qualitätsanforderungen hergestellt werden. Zusätzlich erfordert der komplexe Zusammenhang von Prozessparametern, Filmeigenschaften und Bauteileffizienz eine sorgfältige Optimierung des Herstellungsverfahrens. Ein experimenteller Aufbau muss eine präzise Einstellung der Prozessparameter und Umgebungsbedingungen ermöglichen, aber gleichzeitig die Kompatibilität zu einem großflächigem Beschichtungs- und Trocknungsprozess sicherstellen.

In dieser Arbeit wurde in einem Zwischenschritt von Labor zur Pilotanlage ein technischer Aufbau entwickelt, der Schlitzgussbeschichtung und Prallstrahltrocknung bei inerten oder staubarmen Bedingungen ermöglicht. Sowohl starre als auch flexible Substrate können in dem diskontinuierlichen Verfahren verwendet werden. Alleinstellungs-merkmale sind dabei ein Runddüsentrockner mit lokaler Absaugung, sowie ein Schlitzgusssystem mit hochpräziser Spaltweiteneinstellung und minimiertem Totvolumen. Das Schlitzgusssystem kann ebenso in der Pilotanlage eingesetzt werden. Die Pilotanlage wurde in Zusammenarbeit mit einem Anlagenhersteller entwickelt und im Beschichtungs- und Drucklabor am Campus Nord der KIT aufgebaut. Ein modularer Aufbau zur Vorbehandlung, Beschichtung und Trocknung ermöglicht eine leichte Anpassung an besondere Problemstellungen. Ein System zur Gravur-Beschichtung wurde so ausgelegt, dass entweder fünf unterschiedliche Gravuren gleichzeitig oder nur eine der Gravuren bei einem Totvolumen von 3 ml verwendet werden können. Zur präzisen Auslegung der Beschichtungswerkzeuge und zur Auswertung der

Ergebnisse müssen die fluiddynamischen Eigenschaften der Tinten bekannt sein.

Messdaten für die Viskosität typischer Materialsysteme werden daher vorgestellt und mit empirischen Modellen als Funktion von Zusammensetzung, Temperatur und Scherrate beschrieben. Ebenso wird die Abhängigkeit der Oberflächenspannung von der Lösemittel-zusammensetzung untersucht. Insbesondere für wasserbasierte PEDOT:PSS-Dispersionen kann die hohe Oberflächenenergie der Tinte zur Entnetzung des Films führen. Eine Reduktion der Oberflächenspannung durch Zugabe von Tensiden, sowie eine Erhöhung der Oberflächenenergie des Substrates mittels Plasmabehandlung wird quantitativ beschrieben.

Um die Eignung unterschiedlicher Beschichtungsverfahren zu bewerten, wurden Filme mittels Schleuder-, Rakel-, Sprüh-, und Schlitzgussbeschichtung hergestellt. Dabei wurde dieselbe Tinte und soweit wie möglich identische Prozessbedingungen verwendet. Die Stabilität des Schlitzgussprozesses konnte unabhängig von der Viskosität und Oberflächenspannung der Tinte in einem dimensionslosen Beschichtungs-fenster festgehalten werden. Die trockenen Schichten wurden mittels Rasterkraftmikroskop und UV-Vis-Spektroskopie charakterisiert. Sowohl die Oberflächentopographie als auch die Absorptionsspektren von Filmen, die aus Dispersionen hergestellt wurden, zeigten keinen signifikanten Einfluss des Beschichtungsverfahrens. Im Gegensatz dazu zeigten die aus Polymer-Fulleren-Lösungen mittels Schleuderbeschichtung hergestellten Proben eine stärker ausgeprägte Oberflächentopographie und stärkere Lichtabsorption. Dieses Ergebnis kann durch eine Ausrichtung der Polymerketten aufgrund von Scherkräften in der Beschichtungsströmung erklärt werden. Dieser Effekt kann ein wesentlicher Grund für die niedrigen Wirkungsgrade sein die bei im Pilotmaßstab hergestellten Zellen beobachtet werden. Um diese Hypothese zu überprüfen, wurden unterschiedliche Scherraten mittels Variation der Beschichtungs-geschwindigkeit und der Düsenspaltweite beim Schlitzguss gezielt eingestellt. Die Versuche bestätigten, dass höhere Scherraten eine erhöhte Absorption der photoaktiven Schicht hervorrufen. Besonders deutlich trat dieser Effekt bei dem Einsatz leicht siedender Lösemittel auf. Eine gezielte Erhöhung der Scherintensität in einem technischen Prozess ermöglicht daher eine Verbesserung der optischen Eigenschaften von photoaktiven Filmen.

Die Auswirkungen der Trocknungsgeschwindigkeit auf die optoelektrischen Eigenschaften der photoaktiven Schichten wurden bereits detailliert unter Laborbedingungen untersucht. In dieser Arbeit wurde der Einfluss der Prozessparameter in einem technischen Prozess betrachtet. Unterschiedliche Luftvolumenströme und Temperaturen hatten keinen bzw. nur einen geringen Einfluss auf die Absorption der photoaktiven Schichten. Eine Veränderung der Lösemittelzusammensetzung führte bei höherem Anteil schwersiedender Lösemittel zwar zu einem Anstieg der Absorption, aber auch zu einer Verringerung der Filmstabilität.

Solarzellen mit hybrider photoaktiver Schicht wurden mittels Rakel-, Sprüh- und Schlitzgussverfahren hergestellt und demonstrierten damit die Skalierbarkeit der Technologie. Schlitzgegossene Polymer-Fulleren-Solarzellen wurden mit dem Batch-Coater auf Glas und mit der Pilotanlage auf PET-Substraten hergestellt. Mehrlagige Schichtstapel mit vollständig flüssigphasenprozessierten Elektroden und Anpassungsschichten wurden hergestellt und zeigten vielversprechende Filmbildung und das Potential zur kostengünstigen Herstellung von polymerbasierten Solarzellen.

Content

1. Introduction

1.1. Motivation

Rapidly increasing energy consumption, increasing prices for fossil fuels (Conti and Holtberg 2011) as well as recent changes of energy policies due to the nuclear accident in Fukushima (Joskow and Parsons 2012) create a growing market for low cost renewable energy technologies. The amount of energy that reaches the earth's surface by solar radiation per hour exceeds the annual energy consumption worldwide (National-Geographic-Online 2012). Harvesting and converting solar energy directly by photovoltaic cells may therefore significantly contribute to a clean and sustainable energy supply.

In the last decade, organic semi-conducting and conducting materials were developed which can be processed by solvent based deposition to form functional layers or complete electronic devices (Chen, et al. 2009). By combining layers of conducting and semi-conducting materials, organic light emitting diodes, organic solar cells, and organic field effect transistors can be manufactured. The main advantage of polymer based solar cells (PSCs), compared to conventional inorganic photovoltaics, is their potential to be produced in a continuous roll-to-roll (R2R) process on plastic substrates (Brabec 2008). Attributes such as flexibility, weight, utilization of diffuse light, as well as tunable color, transparency, and shape provide niche markets and possibilities for innovative products. The most relevant figures of merit are efficiency, price and lifetime (Brabec 2004). Whereas lifetime is primarily determined by the chemical stability of the materials and the packaging (Jørgensen, et al. 2008), price and efficiency are significantly influenced by the process which is used to coat and dry the functional layers (Peet, et al. 2009).

At the beginning of this PhD thesis in 2009, Dennler, et al. estimated prices for a rooftop system with Polymer Solar Cells (PSC) between 1.9 and 3 €/Wp[1] and 5 to 50 €-cent/kWh. They concluded that grid parity could be

---

[1] Cost per generated electric power at 1000 W/m² illumination and defined (Air Mass 1.5) light spectrum.

achieved around an efficiency of 7% and a lifetime of 7 years. Shorter life-time and efficiency can be compensated by lower system costs (Dennler, et al. 2009). In the following years the price of conventional photovoltaic systems decreased drastically from consumer prices of ~4 €/Wp in 2009 to ~1.7 €/Wp in 2013 (Solarwirtschaft 2013). Although these values are challenging, PSCs can meet this price range considering current record values for efficiency (9.2% (Service 2011)), lifetime (15 years (Roesch, et al. 2013)) and a price reduction with large volume production (see Figure 1.1). Based on realistic assumptions[2], the pilot-scale coating and drying line that was developed, constructed, and optimized as part of this work has a production capacity of 2.9 million square meters (~420 soccer fields) of polymer solar cells per year. Assuming an average efficiency of 5% these cells could produce a nominal power of 147 MWp, which is equivalent to the electric power of the coal-fired power plant Berlin-Moabit (Vattenfall-AB 2013).

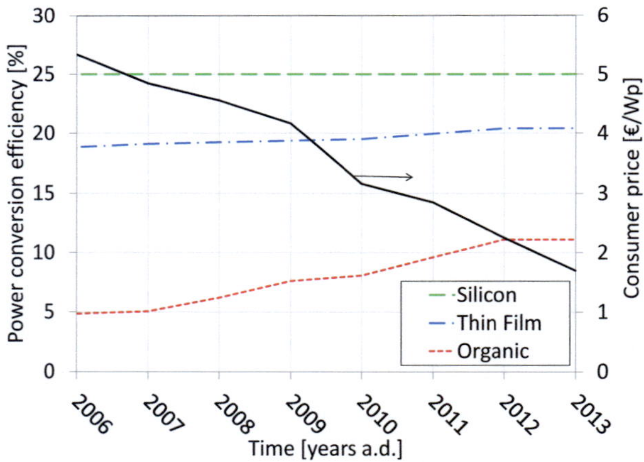

Figure 1.1: Consumer price for rooftop PV-systems (black) (Solarwirtschaft 2013) and record efficiencies of research solar cells for single crystalline Silicon (Silicon; green), Cu(In,Ga)Se₂ thin film (Thin Film; blue), and organic photovoltaic cells (Organic; red) (Wilson 2013).

---

[2] Continuous production at coating speed: 20 m/min, coating width 0.4 m, production efficiency 70%.

Whereas inorganic single junction technologies show no or only moderate increase, organic solar cells have more than doubled their record efficiency since 2006.

Currently the highest published efficiency of a PSC produced in a roll-to roll process is 3.5% (Park, et al. 2010), more than a factor of two below the record efficiencies of laboratory cells (9.2% (Service 2011)). The major challenge is the transfer of a highly sensitive process from controlled laboratory conditions and small areas to a technical environment. Defect free layers of a few nanometers in thickness and lateral dimensions in the order of meters have to be produced (Wengeler and Schabel 2010). Though a number of scientific groups have successfully proven the principle of roll-to-roll manufacturing of PSC (e.g. (Blankenburg, et al. 2009; Krebs 2009c; Kopola, et al. 2010; Park, et al. 2010; Galagan, et al. 2011; Voigt, et al. 2012)) developing a stable process for large area coatings with high average efficiencies and reliability is still one of the major challenges for the technology.

Recently, many European companies manufacturing inorganic solar cells have filed for bankruptcy (Europe 2012). There are, however, still companies developing state of the art organic solar cell technology (eight19, Cambridge, UK; Heliatek, Dresden, Germany; Belectric OPV, Erlangen, Germany), highlighting the need to focus research on this technology in order to sustain a competitive solar industry in Europe.

1.2. Polymer solar cells

In photoactive semiconductors, light is absorbed causing an electron to elevate from the highest occupied molecular orbital (HOMO) to the lowest unoccupied molecular orbital (LUMO). The energy difference between these levels is called band gap and determines the lowest photon energy (or highest wavelength) at which separate charge carriers are created. In contrast to inorganic semiconductors, where the exited electron and the corresponding hole have enough mobility to move independently within the crystal, they remain linked to each other by coulombic forces, creating an electron hole pair (exciton) (Figure 1.2, a). At an interface between a p- and n-type semiconductor (heterojunction), the charges can be separated by transferring the electron from the LUMO of the donor (p-type) to the LU-MO of the acceptor (n-type) if the energy difference between the LUMOs is larger than the exciton binding energy (Hou, et al. 2008).

1.2.1. Working principle

The separated charges are subsequently transported through the respective semiconductor toward the electrodes where they provide an electric potential for an external circuit. The difference between $LUMO_{Acceptor}$ and $HOMO_{Donor}$ is proportional to the open circuit voltage (V_{OC}) and thus directly influences the potential of the cell (Brabec, et al. 2001). Electrons are extracted at the electrodes, if the energy level of the anode is lower than the $LUMO_{Acceptor}$, and if the energy level of the cathode is higher than the $HOMO_{Donor}$ (Brabec 2004).

Material systems for polymer solar cells are fine-tuned for highest efficiency, hence small changes in energy level can have a strong impact on device performance (Scharber, et al. 2006). Contamination, residual solvent, absorbed water, or oxygen may change the energy levels and subsequently the behavior of the solar cell. These interactions have to be considered in the development of a technical process and care has to be taken when choosing additives or handling the material under ambient conditions.

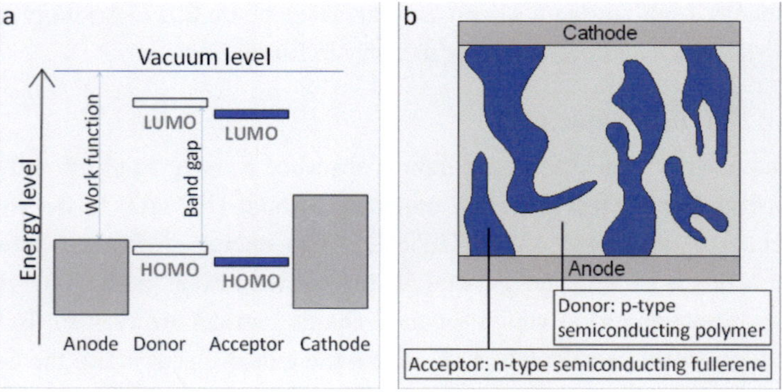

Figure 1.2: Schematic sketch of energfigurefigurey levels of electrodes, donor and acceptor during vacuum level alignment (a) and a sectional view of a bulk-heterojunction structure (b).

To facilitate the transport of charges, materials with different work function[3] are used as electrodes. Connected in an electric circuit, the energy level of their conductive bands approach (or align for a short circuit) and thus shift the vacuum level at the interface of the photovoltaic cell, creating an electric field. If the work function of the anode is lower than that of the cathode, electrons are driven towards the anode whereas holes move towards the cathode (see Figure 1.2, b) (see e.g. (Brabec 2008; Colsmann 2008)).

The average distance an exciton diffuses before recombining is only 5-10 nm whereas film thicknesses of more than 100 nm are required to absorb a large fraction of the solar radiation (Lunt, et al. 2009). When a solution of a semi-conducting polymer (p-type) and fullerene (n-type) blend is coated and dried, polymers and fullerenes phase separate into crystalline and amorphous domains. This interpenetrating network of p- and n-type domains is called bulk-heterojunction (BHJ) and creates a large pn-interfacial area throughout the bulk layer. The morphology of the BHJ is defined by degree of crystallinity, domain size, and crystal orientation. An optimized morphology is achieved if the domain size is in the order of the exciton diffusion length (reducing recombination), while providing direct pathways to the electrodes in both phases (reducing transport losses) (Shaheen, et al. 2001a). The structural evolution during processing and subsequently the final morphology of the film depend on thermodynamic material properties as well as kinetic process parameters (Ma, et al. 2005; Sanyal, et al. 2011a; Sanyal, et al. 2011b).

1.2.2. Electric performance and characterization

The electric performance of a solar cell is characterized by the current-density (j) voltage (V) behavior under defined illumination. Without illumination, a solar cell shows high resistance at low voltages and rapidly increasing currents with increasing voltage (Figure 1.3.a). This characteristic is shifted into the lower right quadrant under illumination. For every operating point within this quadrant, a positive voltage correlates to a negative

---

[3] Energy difference between a free electron and electrons in the conducting band.

current density, thus electric energy can be consumed in an external circuit. The intersection with the current density axis specifies the short current density (j_{SC}) whereas the intersection with the voltage axis specifies the open circuit voltage (V_{OC}). The point at which the product of voltage and current reaches a maximum (MPP= maximum power point) can be expressed as a function of j_{SC} and V_{OC} by introducing the Fill Factor (FF):

$$FF = \frac{j_{MPP}V_{MPP}}{j_{SC}V_{OC}}.$$

(1.1)

The efficiency of the solar cell is then defined as the electric power density (P_{MPP}) at MPP, divided by the incident solar power density (i_{solar}).

$$PCE = \frac{P_{MPP}}{i_{solar}} = \frac{FF \cdot V_{OC} \cdot j_{SC}}{i_{solar}}$$

(1.2)

For measurement under an artificial light source, a mismatch factor accounts for the difference compared to the solar radiation spectrum (Brabec 2004).

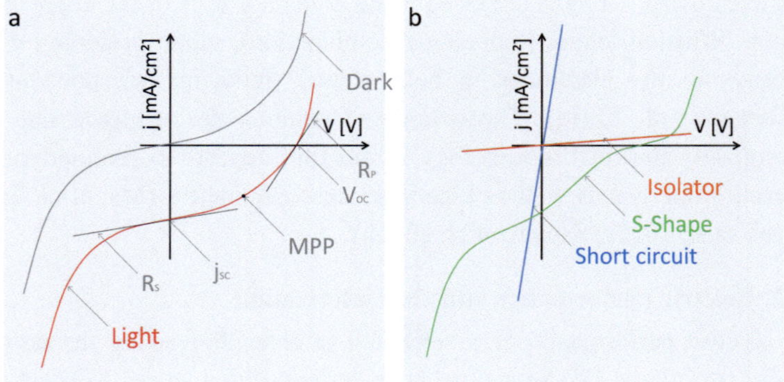

Figure 1.3: Sketch of characteristic current density – voltage curves. (a) Current density voltage curve of a working solar cell with (red) and without illumination (grey). (b) Characteristic curve of a short circuited cell (blue), an isolated cell (orange) and a cell with accumulated charges at an interface (green).

Current-density voltage curves can be used to identify possible reasons for lower performance of solar cells. The short circuit current density is a measure of the mobility and lifetime of charges and may indicate an unfavorable morphology. The open circuit voltage is primarily affected by the energy levels of the semi-conducting and electrode materials. The reciprocal slope at V=0 and j=0 can be simulated with an equivalent circuit as a series (R_S) or parallel (R_P) resistance, respectively. A high series resistance indicates ohmic losses in the electrodes whereas a low parallel resistance results from leakage currents through the BHJ layer (Brabec 2004). Figure 1.3.b shows characteristic curves commonly observed for cells with low efficiency. A linear curve with large slope represents a cell with a short circuit from one electrode to the other (blue), e.g. due to pinholes or particulates in the semi-conducting film. Linear curves with a small slope are caused by cells that act as isolator (orange), e.g. due to defects in the electrodes or insufficient pathways in the BHJ-layer. An "S-shape" behavior (green) is seen in cells where charges accumulate at one of the interfaces (Wang, et al. 2011), e.g. due to oxidation of the cathode or surface defects.

1.2.3. Device architecture

The photoactive layer is sandwiched between a thin (<50 nm) hole blocking layer at the cathode and an electron blocking layer at the anode. These adaptation layers adjust the energy level of the electrodes and inhibit short circuits through the cell. If one of the electrode layers already exhibits a suitable energy level, only one of the blocking layers is needed. The generated current is transported by an anode and a cathode layer to external contacts or connected cells (see Figure 1.4). At least one of the two electrode layers has to be transparent so that solar radiation can reach the active layer. The other electrode can be transparent as well, creating a semi-transparent cell (Colsmann, et al. 2011), but more commonly a reflective back electrode that reflects the light back into the photoactive layer is used. Depending on whether the cathode is at the top or bottom of this stack, it is referred to as "standard" or "inverted" architecture. In standard cells the sunlight often passes through a transparent substrate, a so called superstrate configuration.

To connect several cells in series, an additional patterning step or structured coating is necessary. External contacts or bus bars can be applied by established printing technologies (e.g. screen printing (Krebs 2009c) or inkjet (Galagan, et al. 2012)).

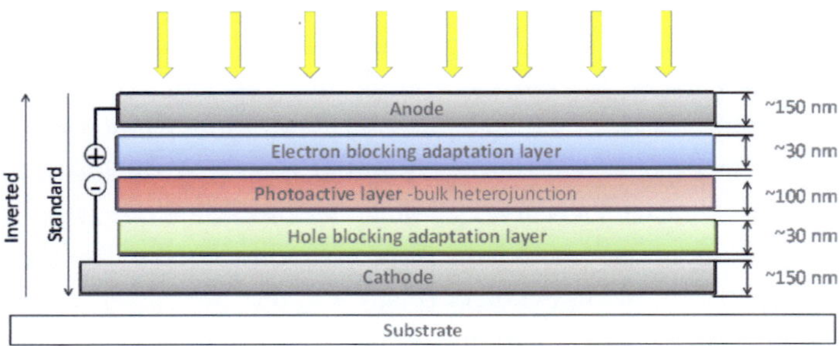

Figure 1.4: Schematic sketch of the layers in an inverted polymer solar cell. The photoactive layer is sandwiched between a hole-blocking and a cathode layer on one side and an electron-blocking and an anode layer on the other side.

Figure 1.4 shows typical film thicknesses for the individual layers in an inverted polymer solar cell structure. The precise thickness of each layer has to be determined for each material combination and module architecture. A thicker photoactive layer for example, increases light absorption as well as transport losses. The required electrode thickness is determined by the local current density and the distance to the external contacts, bus bars, or connected cells. As this stack of semi-transparent layers has a thickness in the same order of magnitude as the optical wavelength of light, wave interference occurs between incident and reflected light and may cause constructive or destructive interference within the photoactive layer.

Due to the strong interactions between film thickness, material properties, and performance, it becomes evident that film homogeneity and precise control of process parameters are key factors for a successful scale-up. As well, material properties may change due to different process conditions, making a pilot-scale investigation and validation of lab scale experiments inevitable.

1.3. From laboratory to pilot-scale

1.3.1. Material classes and nomenclature

The term "organic" in OPV refers to the composition of the photoactive layer only. Other layers can consist of organic, in-organic, or hybrid material systems, including polymers, small molecules, metals, metal oxides, and salts. Most layers can be prepared by either gas or liquid phase deposition.

Barrier layers with water vapor transmission rates below 10^{-3} g/m²d are required for encapsulation (Cros, et al. 2011) and are typically manufactured by gas phase sputter deposition (Dennler, et al. 2006). *Electrodes* based on metals (e.g. Aluminum (Al) or Calcium (Ca)) or transparent conductive oxides (e.g. Indium tin oxide (ITO)) are often applied by physical vapor deposition or sputtering, respectively. However, materials for liquid phase deposition of transparent polymer electrodes (e.g. poly(3,4-ethylenedioxythiophene) poly(styrenesulfonate) (PEDOT:PSS)) or opaque metal electrodes (e.g. silver (Ag) nanoparticles) are readily available and can perform as well as evaporated systems at lower process costs (see e.g. (Kim, et al. 2011)). *Adaptation layers* are often composed of metal salts, metal oxides (e.g. Zinc oxide (ZnO)) or polymers (e.g. PEDOT:PSS) and can be coated by liquid or gas phase deposition (see e.g. (Colsmann 2008)). *The photoactive layer* of an organic solar cell can be composed of small molecules, polymers, or a blend of both. Whereas polymers and polymer blends are commonly applied by liquid phase deposition, small molecule solar cells can also be evaporated (see e.g. (Lin, et al. 2012)).

In this work, the term *"polymer solar cell"* (PSC) will be used for cells with liquid phase deposited active layer, bulk-heterojunction architecture, a p-type polymer as electron donor, and a fullerene derivative as electron acceptor. A typical model system, that has been studied extensively in literature is the polythiophene polymer Poly(3-hexylthiophene) (P3HT) and the C60-fullerene derivative Phenyl-C61-butyric-acid-methyl-ester (PCBM). In *"hybrid solar cells"* (HSC), PCBM is replaced by semi-conducting nanoparticles. These so called quantum dots (QD) exhibit higher absorbance, higher mobility compared to PCBM and opto-electric properties can be adjusted by tuning size and shape of the particles (Kumar and Scholes 2008).

1.3.2. The laboratory scale process

On laboratory scale, organic and hybrid solar cells are conventionally produced by spin coating of solutions or dispersions on ITO coated glass substrates. The ink is dosed on a substrate which then starts to rotate, distributing the ink homogeneously on the surface due to centrifugal forces. The sample is rotated continuously until it is dry. The final thickness is controlled by spin frequency, composition, material properties of the ink, and drying conditions. For a given setup and material system this method yields reproducible coatings with small amounts of coating fluid (Kistler and Schweizer 1997). However, the spin coating process parameters are strongly interdependent. A change in spin frequency results in a lower thickness but also in a higher drying rate. A change in composition of the applied fluid causes different dry film thicknesses as well as drying rates. The design of the experimental setup also influences the film formation as acceleration and air flow may differ. A second problem is the local inhomogeneity of process parameters on a spinning sample. Drying rate and hydrodynamic shear-rate are functions of the distance from the rotational axis.

Spin coating is limited to small rigid substrates and a scale-up to a technical and continuous roll-to-roll process is very difficult due to many unknown and unspecified parameters. Processing conditions determine the surface homogeneity and morphology of the film which in turn have a strong impact on cell efficiency. Thus, a given material system that was optimized by lab scale spin coating might produce completely different results on a technical scale. As a technology aimed at lowest cost rather than highest efficiency, the transfer of the production process to a technical scale process is mandatory for the success of PSCs.

1.3.3. Roll-to-roll processing

As discussed by Krebs and Søndergaard (Krebs 2009a; Søndergaard, et al. 2012a; Søndergaard, et al. 2013), in principle a variety of methods is suitable for coating of functional films in polymer solar cells. Table 1.1 gives an exemplary overview of cell efficiencies achieved by various coating methods and processing conditions. Spray coating and doctor blading have produced cells with efficiencies close to those of spin coated samples. However, these cells were produced on small (<25 mm^2) ITO coated glass substrates under inert conditions. A remarkable efficiency of 3.5 % was achieved by Park et al. with a modified roll coating method, indicating that a scale-up without efficiency loss is possible. Park covered the photoactive film with a semipermeable silicon sheet after coating, thereby reducing the drying rate, changing the surface energy at the top interface and introducing viscous shear. Though no average efficiency, device area and atmospheric condition are specified, the roll-to-roll samples performed better than the spin coated reference devices in this study (Park, et al. 2010). Promising results by Kopola et al. (Kopola, et al. 2010), Zimmermann, et al. (Zimmermann, et al. 2011), Blankenburg et al. (Blankenburg, et al. 2009), Schrödner, et al. (Schrödner, et al. 2012), and Angmo et al. (Angmo, et al. 2012) show that gravure and slot-die coating are technologies with high potential.

Table 1.1: Incomprehensive comparison of various processes used for polymer solar cells based on P3HT:PCBM.

Coating method (active layer)	Substrate	R2R	Coated electrodes	Active area [mm^2]	PCE [%]	Ref
Spin coating	Glass	No	0	4	5.0	(Irwin, et al. 2008)
Spray	Glass	No	0	25	4.1	(Susanna, et al. 2011)
Doctor blade	Glass	No	0	16	4.0	(Chang, et al. 2009)
Ink-Jet	Glass	No	0	9	3.7	(Eom, et al. 2010)
Roll	PET	Yes	0	7	3.5	(Park, et al. 2010)
Brush painting	Glass	No	0	4	3.3	(Kim, et al. 2007b)
Gravure	PET	Yes	0	19	2.8	(Kopola, et al. 2010)
Slot-die	PET	No	0	1320	2.2	(Zimmermann, et al. 2011)
Slot-die	PET	Yes	0	25	1.7	(Blankenburg, et al. 2009)
Spin coating	Glass	No	2	300	1.6	(Angmo, et al. 2012)
Slot-die	PET	No	1	100	1.5	(Angmo, et al. 2012)
Slot-die	PET	Yes	2	3550	0.4	(Angmo, et al. 2012)

Table 1.1 illustrates the "red bricks" for a successful scale-up from lab scale spin coating to a technical process: the transfer to another coating method, flexible substrate, ambient R2R processing, solvent based deposition of electrodes and adaptation layers, and large active areas. Due to the process sensitive nature of the materials, each of these steps can lead to a decrease in power conversion efficiency and requires optimization.

Cell architectures, such as the one illustrated in Figure 1.4, can be prepared using vacuum free coating and drying processes only. The individual layers can be applied subsequently within one continuous roll-to-roll process. In general, each layer is applied in four steps:

- Pre-treatment (e.g. cleaning, surface activation, …),
- coating,
- drying,
- and post-treatment (e.g. annealing, sintering, …),

though pre- and post-treatment may not be required for every layer. From a process technological viewpoint, coating and drying can be seen as unit operations. In the first step a coating ink is distributed in cross web direction and forms a thin "wet film"[4] on the desired substrate. This film is subsequently dried to form a layer with certain optical and electrical properties.

Inherent difficulties of R2R production of polymer solar cells are flexible encapsulation, stability under ambient conditions, and contamination of the functional materials. The large holdup of conventional coating tools is a practical barrier for scarce, newly-developed materials.

On the other hand, most large scale processes allow for a better optimization of the unit operations, coating and drying, because process parameters can be varied independently. A continuous steady state process also eliminates instabilities and material loss due to the transient coating flow at the

---

[4] The term "wet film" is used to describe a solvent based film after deposition, independent of whether the solvent is an organic solvent or water.

start and end of a batch (or sheet by sheet) coating process and may have a material efficiency[5] close to 100%.

1.3.4. Impact of process parameters

Though polymer solar cells have been prepared by a variety of coating methods (see Table 1.1), a direct comparison and evaluation is not possible as different materials and processing conditions are used by different scientific groups (Vak, et al. 2007; Blankenburg, et al. 2009; Chang, et al. 2009; Hoth, et al. 2009a; Krebs 2009b; Kopola, et al. 2010; Park, et al. 2010; Galagan, et al. 2012; Voigt, et al. 2012). Since materials for PSCs are still expensive, the research community has evaded a systematic investigation of process parameters on pilot-scale (Søndergaard, et al. 2012b). Scientific papers on pilot-scale coating typically test a process configuration and show a proof of principle (Søndergaard, et al. 2013). A variation of an independent parameter in a pilot-scale coating process has not been published to the best of my knowledge.

The influence of drying rate (Li, et al. 2005; Mihailetchi, et al. 2006; Schmidt-Hansberg, et al. 2009b; Schmidt-Hansberg, et al. 2011c) and annealing conditions (Kim, et al. 2007a) has been investigated thoroughly on laboratory scale. At slower evaporation rates and lower temperatures the phase segregations and crystallization is enhanced up to a critical drying time where the BHJ is assumed to have reached an optimized structure and no further improvement is possible. Annealing under elevated temperature or an atmosphere containing a solvent can further improve performance, especially for processes with fast drying rates (Schmidt-Hansberg, et al. 2011b). These results were obtained on a small scale knife coating set-up and have not been verified under technical conditions. Several publications have reported that solvent composition has a strong impact on film homogeneity and performance in pilot-scale processes (Kopola, et al. 2010; Schrödner, et al. 2012; Voigt, et al. 2012), stressing the need for further studies.

---

[5] Material amount in the film, divided by the amount used in the manufacturing.

1.4. Aim and scope of this work

The technical processing of (semi-)conducting films from inks thus is one of the most important factors for the success of polymer solar cells. A strong interaction between process parameters, film morphology, and film properties results in a need for pilot-scale research at an early point in the development.

Compared to conventional coatings (e.g. color coating) the functional films in polymer solar cells exhibit electrical and optical properties (e.g. light absorption, conductivity, work function) in addition to film quality (homogeneity and number of defects per area). As process parameters may affect film quality as well as film properties, an optimization needs to address both aspects.

In summary the aim of this work is to

- establish a technical process for coating and drying of functional films,
- investigate the impact of coating methods and process parameters,
- and validate the influence of drying kinetics under technical conditions.

To address these points, an experimental set-up and procedure for the manufacture of functional films must be designed first (Chapter 2). An intermediate scale coating and drying set-up will bridge the gap between laboratory and technical roll-to-roll processing. This "batch-coating" set-up shall feature precise control of process parameters while employing technical tools for coating and drying. In cooperation with a roll-to-roll plant manufacturer a pilot-scale coating line was developed, incorporating all basic components for a coating and drying sequence. Eventually, a slot-die and gravure coating system were customized in order to fulfill the requirements for the production of functional films for PSCs.

The experimental chapter of this work (Chapter 3) specifies the materials, procedures and characterization methods. Experimental details, such as ink composition, mixing procedure, measurement protocol, and substrate treatment, strongly influence the experimental results and have to be recorded thoroughly in order to get reproducible results.

Material properties, such as viscosity, surface tension, and contact angle, determine the coating flow. The design of coating tools, dimensionless description of experimental results, and their numerical simulation are based on experimental data or approximate models for these properties. Since many of the materials in PSCs have been developed recently and have never been applied in a technical process before, data on their fluid-dynamic properties are scarce. The solid content and material type affect the shear thinning behavior of coating inks. As values for the viscosity of the inks are typically less than 50 mPas, the solvent may contribute a significant part of the overall viscosity. The contact angle of a droplet is determined by the liquid's surface tension and the substrate's surface energy. Both may have to be adjusted to allow for a good wetting behavior of aqueous solutions and dispersions. Experimental data and predictive models for typical coating inks, substrates and films are presented in Chapter 4.

An evaluation of coating methods and the impact of process parameters are shown in Chapter 5. From the large set of possible coating methods, the most promising candidates shall be identified, based on material properties and desired operating conditions. For self-metered processes the thickness is a function of operating parameters and has to be calibrated. The dependence of thickness and homogeneity on process parameters limits the process to regions with sufficient homogeneity. Slot-die coating, a pre-metered coating method, does not require a thickness calibration. Its stability limit and application will be discussed in detail. Furthermore, film properties may depend on the position of the operating point within a stable operating or coating window. A variation of coating velocity and coating shear-rate will be discussed with respect to the optical and electrical characteristics of the coated film.

The scale-up of the drying process must address two fundamental issues: the local distribution of transfer coefficients, and the impact of drying kinetics on film stability and evolution of the BHJ (Chapter 6). In lab scale processes, a discrete substrate is coated under conditions where the local drying rate is a function of the position on the substrate, resulting in a local variation of film properties. If only a small fraction of the coated area at a fixed position is used, this may be acceptable for research purposes. In a production processes however, the complete coated area has to be dried under identical conditions to ensure homogeneous product quality. Typical values for heat transfer coefficients (HTCs) in technical impinging jet driers range from ~10 to ~500 W/m²K whereas the impact of drying kinetics has been investigated at HTC values below 10 W/m²K. A low drying rate typically improves the morphology of the BHJ but may allow a de-wetting process to take place. Problems related to de-wetting of the film are suppressed during laboratory investigations by the use of plasma treated glass or silicon substrates. Thus, a trade-off between optimized morphology and processability in a pilot-scale process will be proposed.

Though no optimization of power conversion efficiencies could be realized within the timeframe of this work, it is important to demonstrate whether working solar cells can be produced with the set-up, materials, and processes discussed before (Chapter 0). Photoactive layers composed of polymer-nanoparticles and polymer-fullerene blends were processed using the batch and pilot-scale set-up. Finally, multilayer stacks for standard and inverted geometries were prepared including fully coated electrode and adaptation layer systems. The discussion of these experiments marks the state of the process at the end of this thesis and points to routes for further optimization.

2. Design and implementation of experimental set-up

In the last decade, transfer to other coating methods has been done by applying conventional coating tools for the deposition of functional films in PSCs. Using tools and equipment designed for other purposes often leads to operating points outside the optimum range. An addition of additives to change material properties such as viscosity and surface tension is often not possible because the film has to keep its original electrical and optical properties. Instead of adjusting the ink to the requirement of the process, which is the canonic procedure for complex processes in the graphic printing industry (e.g. offset-, or flexo-printing), the experimental set-up and coating tools need to be customized to the requirements of the new technology.

2.1. Introduction and state of the art

In 2001, Saheen, et al. (Shaheen, et al. 2001b) published the first paper on solar cells with screen printed active layer. Krebs, et al. demonstrated large area screen printing of complete solar cells using industrial equipment (Krebs, et al. 2009). Even though he employed a sheet-to-sheet printer, the holdup of the tools required more than 100 ml of ink for each layer. First roll-to-roll slot-die coating of a photoactive layer was reported by Blankenburg, et al. in January of 2009 (Blankenburg, et al. 2009).

Recently, several approaches to customize experimental set-up and coating tools were proposed. Kopola, et al. (Kopola, et al. 2010) and Voigt, et al.(Voigt, et al. 2012) tested different gravure patterns for gravure printing of photoactive and hole injection layers. Krebs used a "meniscus guide", a flexible protruding metal foil, in slot-die coating to minimize the need for high precision positioning of the coating gap (Krebs 2009b). A complete roll-to-roll line was set up in a cleanroom environment at the Holst Centre in Eindhoven (Galagan, et al. 2011). In 2012, Dam, et al. presented a slot-die with minimized hold-up in order to conduct experiments with newly developed materials.

In contrast, very little attention has been given to the development of customized drying tools. In laboratories, films are typically dried by natural or forced convection. Under most conditions a drying front moves over a stationary substrate. A drying front is a line separating dry and wet areas, defined by a threshold concentration. Due to a discontinuous change of optical properties at low solvent content, a clearly defined line can be seen on the drying film. During spin coating or natural convection on a hot-plate, faster drying rates are observed at the edges compared to the center of the film. A laminar flow over a stationary plate induces decreasing drying rate in flow direction (Schmidt-Hansberg, et al. 2011a). On technical scale, overall homogeneity may increase, as the film moves through the drier averaging local inhomogeneity in web direction. However, impinging jets of drying-air still induce a pattern of different transfer coefficients, which is characteristic for a nozzle geometry and volume flow rate and needs to be optimized with respect to homogeneity (Cavadini, et al. 2012a).

At the start of this work, the experimental equipment present in the laboratory consisted of a Zehntner linear drive and a knife coating set-up in a laminar-flow drying channel. Though this set-up allows for precisely defined drying conditions, differences in local drying speed have to be considered and the integral heat transfer coefficients are limited to values below 10 W/m²K (Schmidt-Hansberg 2012). Due to the lack of a solvent exhaust, experiments with technical coating equipment were not possible.

2.2. Technical laboratory coater (BC)

To facilitate the scale up from highly defined laboratory conditions to pilot-scale operation, a technical laboratory coater was designed in collaboration with Katharina Peters. This intermediate step should feature a precise control of ambient conditions while allowing the application of technical tools for coating and drying.

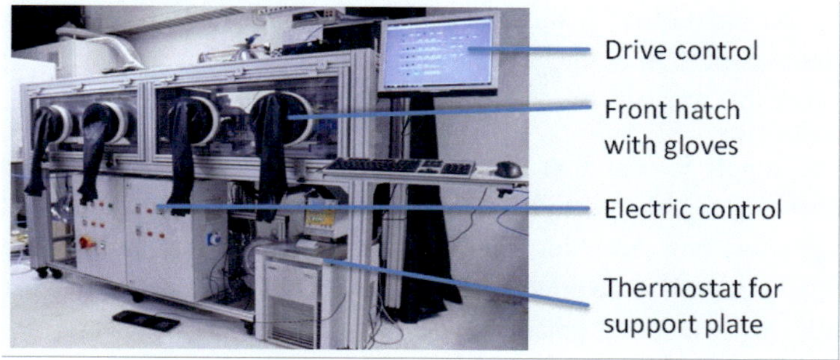

Figure 2.1: Photograph of the encapsulated coating chamber of the batch-coating and drying set-up.

2.2.1. Control of environmental conditions

Oxygen, humidity, and dust particles may compromise the functionality of semi-conducting films. In addition, large amount of solvents may evaporate when using technical coating tools. The set-up should thus provide:

- control of oxygen and humidity level,
- low dust particle concentration,
- and a solvent exhaust or solvent separation.

The control of oxygen and humidity level was achieved by building an encapsulated chamber that can be flushed with nitrogen (**inert operation**). The tools and materials inside the encapsulation can be operated through gloves. A concentration of less than 1 mol.-% oxygen and water was measured after 20 min flushing with nitrogen. Inert operation is not always necessary and complicates the handling. During **open operation**, the nitrogen flow is replaced by a blower providing filtered air, while the coating chamber can be accessed through a large front hatch (see Figure 2.1).

The gas flow into the chamber is distributed at the top through a perforated metal sheet and particle filter mats. The exhaust is positioned at the bottom, resulting in a laminar downward flow with reduced dust particle concentration. After ~12 min of inert operation, particle concentration corresponding to a clean room class[6] 6 was reached after 12 min. A detailed description and characterization of the gas management system is documented in a project work that was supervised together with Peters (Hotz 2012).

The air supply for the drier was first set up in a closed circulation over a heat exchanger and an active-coal solvent filter. This configuration led to a high pressure drop, resulting in a limited volume flow rate. As well the active coal filter had to be dried for more than 2 hours prior to inert operation, in order to reduce the amount of absorbed water. Subsequently, an improved flow configuration was designed in collaboration and implemented by Scharfer and Jaiser.

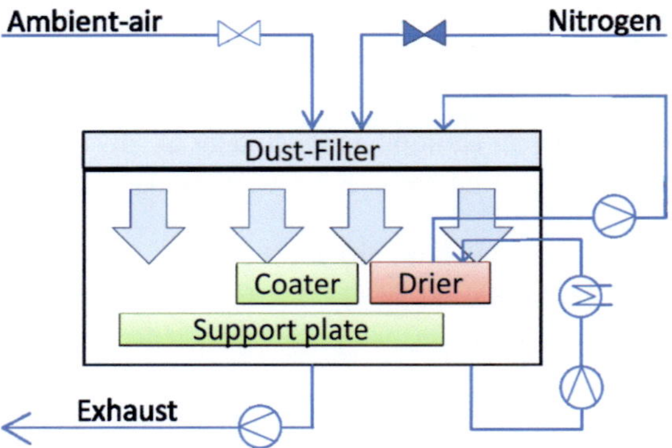

Figure 2.2: Schematic flow diagram of the gas management system of the Batch-Coating-System.

The schematic flow diagram of the final configuration is shown in Figure 2.2. Instead of a close loop over the drier, individual blowers supply air from the bottom of the coating chamber directly to the drier and extract exhaust air from the drier which is fed back into the coating chamber at the top. In this configuration, the solvent, water, and oxygen concentrations are reduced by the flow of nitrogen through the system.

---

[6] ISO-Norm 14644-1 (equivalent to class 1000 according to US Federal Standard 209E)

The coating and drying equipment, indicated in Figure 2.2 by a green and red box, respectively, are described in the next subsections.

2.2.2. Slot-die coating system

The first step towards pilot-scale was employing a fixed coating tool and a moving substrate. The substrate is supported by a glass covered, aluminum plate which is temperature controlled by a thermostat (Proline RP 855, Lauda). The support is mounted on two sliding carriages on a linear positioning system (TR24x5, ELT60, Lineartechnik Korb) with a total length of 2200 mm. The positioning system is driven by a step motor (ST8918S4508, Nanotec) with velocities of up to 30 m/min (=500 mm/s). The calibration and minimization of vibrations was investigated by Hotz (Hotz 2012) and is summarized by Peters (Peters 2013).

Coating can be done with a coating knife (ZAA2000, Zehntner), facilitating scale-up from laboratory equipment, or with a specially designed slot-die coating system, facilitating scale-up towards pilot scale.

The key requirements for the slot-die coating system and technical means to address them are listed in Table 2.1.

Table 2.1: Summary of the most important requirements for the slot-die coating system and their technical solution.

Requirement	Technical solution
Low hold-up	Specialized design of coating die cavity, reduced coating width
Precise gap adjustment	Relative positioning by runner system
Facilitate scale up	Identical coating tool for batch and roll-to-roll processing
Easy handling (start-stop, venting, cleaning)	Floating bearing of slot-die

Low hold-up is essential as material prices are still high. Preparing 100 ml ink for the photoactive layer, with a solid content of 12.5 mg/ml P3HT and

12.5 mg/ml PCBM would require material worth 846 € assuming current prices at Sigma-Aldrich (Sigma-Aldrich 2013). More advanced materials, such as low band-gap polymers, or semi-conducting nanoparticles are even more expensive and may not be available in sufficient quantities. To reduce the hold-up, the coating die cavity was optimized for the viscosity and flow rates expected in organic electronic solutions (see Appendix 9.3.1), the coating width was reduced to ≤ 50 mm and a syringe pump was installed close to the coating die.

Small dimensions and weight of the developed slot-die are a necessary pre-condition for the technical solutions to the following points in Table 2.1.

In a conventional configuration, the slot-die is positioned in a defined absolute distance from the rotational axis of the back-up roll which supports the substrate. The clearance of the bearing (>3 µm), the symmetry of the back-up roll (>3 µm), and the thickness variation of the substrate (>4 µm) contribute to an error of more than 10 micrometers in gap width[7] (Döll 2013). The width of the capillary gap determines the stability of the coating process (see Subsection 5.3.5). Numerical simulations by Durst and Wagner (Durst and Wagner 1997) showed that the gap width should be in the same order of magnitude as the wet film thickness to inhibit the formation of large eddies and stagnation zones. For films in polymer solar cells which often require wet film thicknesses of less than 10 µm, a gap in the same order of magnitude would result in large relative variation in gap width. Whereas a variation of e.g. 10 µm may be acceptable for a coating with a fixed gap of 200 µm, it would result in a relative error of 25% for a gap of only 40 µm. To achieve a higher precision in gap width, the die was mounted on a runner system to allow for a relative adjustment with micrometer precision. In the first design, the die was positioned using micrometer calipers, and then fixed in position with an Allen screw. The clearance of the bearings in combination with the torque during fixing resulted in a small displacement of the die. This problem was eliminated in the second design, which used linear bearings as guide (see Figure 2.3).

---

[7] The "gap width" is defined as the distance between the die's lip and the substrates surface.

The positioning was realized by applying a force through the slot-die-frame, onto micrometer calipers. The micrometer calipers are permanently fixed to the slot-die frame and push against the support frame, positioning the two frames relative to each other. As the support frame slides directly on the substrate, a relative positioning of die lip and the substrate's surface is achieved.

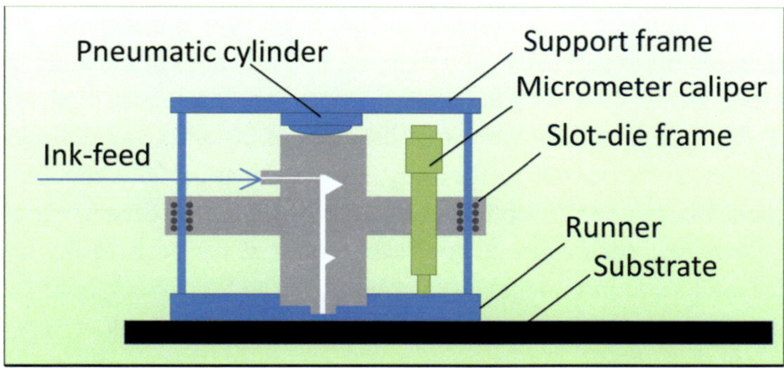

Figure 2.3: Schematic sketch of the slot-die support. The slot-die frame (grey) is positioned relative to the support frame (blue) by micrometer calipers (green).

This "2nd generation" slot-die system was designed and developed in collaboration with Schmitt, who elaborated the design and applied it to other coating applications (Schmitt, et al. 2013).

2.2.3. Technical drying

One of the most common forms of technical driers for functional films is an impinging-jet drying unit. Driers using solely infra-red radiation (IR-drier) or heat conduction (contact drier) increase the temperature throughout the film. In contrast, impinging-jet driers additionally increase the mass transfer coefficients (MTC) as well as the heat transfer coefficient (HTC) into the surrounding drying gas. In the drier of a fast R2R process, a high level of homogeneity in transverse machine direction (TD) has more relevance than homogeneity in machine direction (MD) because inhomogeneity in MD can be averaged by the fast moving substrate (web). In case of a product that requires lower web velocities, the influence of the homogeneity of the transfer coefficients in MD increases. As a limiting case a stationary substrate may be considered. Here, a high level of homogeneity of the distribution of the transfer coefficients in both lateral directions has to be ensured to guarantee a uniform quality in functional films.

Typically, drier nozzles are arranged either as array of round-nozzles (ARN) or as array of slot-nozzles (ASN) (Martin 2002). Inhomogeneity in a confined impinging jet system can occur on two dimensional scales. **Large scale inhomogeneity**, of a dimension in the order of the array width or length, can be caused by the cross flow effect (Florschuetz, et al. 1981). The spent fluid leaves the domain to the sides of the array and deflects impinging jets at the outside of the array. Hence, the cross flow effect results in higher transfer coefficients at the center of the array. Stagnating air or rigid walls may also cause lower heat transfer at the edge of the nozzle field. **Small scale inhomogeneity** is caused by the flow structure of an impinging jet itself. In the stagnation region of an impinging jet, a very thin boundary layer forms and grows in radial direction. This results in a local distribution with a local maximum of the heat transfer coefficient below each jet of the array.

Array of round nozzle (ARN-) drier

To realize homogeneous drying of functional films on stationary substrates, a drier concept with an array of round nozzles was implemented first. To counter large scale inhomogeneity, local outlet nozzles were spaced evenly throughout the array (Figure 2.4). Inlet and outlet nozzles are connected to individual plenum chambers, distributing the air flow through all nozzles.

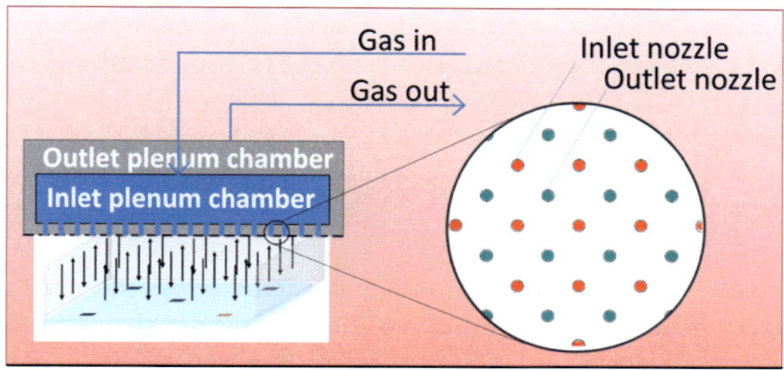

Figure 2.4: Schematic sketch of the array of round nozzle drier designed for drying functional films on stationary substrates.

By choosing a small nozzle spacing (L_T=28 mm) and a large distance from nozzle to substrate (H_{NSD}=100 mm), the formation of a well-defined stagnation point was inhibited. Figure 2.5 shows the temperature field on the surface of a plate heated with the ARN-drier with different nozzle to substrate distances. At H_{NSD}=20 mm the small scale inhomogeneity causes an inhomogeneous temperature distribution; each nozzle is clearly visible. At H_{NSD}=100 mm, the jets of impinging air merge and a homogeneous temperature distribution is achieved at the center of the array.

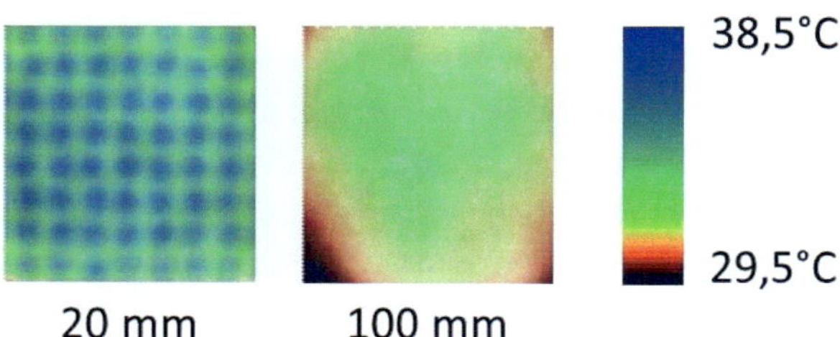

Figure 2.5: Temperature distribution on the surface of a plate heated by the ARN drier. For a nozzle to web distance of 20 mm, higher heat transfer coefficients under each nozzle results in an inhomogeneous temperature field.

Though the temperature distribution indicates a homogeneous heat transfer at technical drying rates, a direct scale-up is not evident. At 100 mm the ratio of H_{NSD} to L_T is much larger than the value proposed for an energy efficient drier (Martin 2002):

$$\frac{H_{NSD}}{L_T} = 3.6 \gg \frac{H_{NSD,opt}}{L_T} = \frac{5}{7}. \tag{2.1}$$

With increasing H_{NSD} the overall energy efficiency of the drier decreases, because a large fraction of power used in the blower is dissipated in the overlapping impinging jets without affecting the boundary layer thickness. The search for a nozzle geometry, optimized for homogeneity as well as energy efficiency, was continued by Cavadini (Cavadini, et al. 2012b) and led to a patented nozzle design (Cavadini, et al. 2013a).

Heat transfer coefficients of up to 60 W/m²K were realized with this set-up (H_{NSD}=100). The HTC calculated from empirical correlations showed good agreement with measured values. A more detailed characterization is presented in Appendix 9.5.1. Temperature, velocity and pressure of the drying air were measured in-situ, using a heat-wire anemometer and an analog manometer. With these values the HTC was calculated for every experimental setting.

Array of slot nozzle (ASN-) drier

A conventional array-of-slot-nozzles-drier was installed by Scharfer and Jaiser to realize higher transfer coefficients and direct scalability. The ASN configuration, with slots aligned in cross web direction, is often applied in roll-to-roll driers. If designed to inhibit cross flow, it creates a field of local transfer coefficients with symmetry to the center of the web and periodic variations in web directions. Thus, for stationary or slow moving substrates this configuration leads to periodic inhomogeneity. Heat and mass transfer are highest close to the stagnation line below the nozzle and decreases with increasing distance from the stagnation region. To emulate the conditions in a technical drier, the support plate of the batch coater was moved back and forth inside the ASN drier.

2.2.4. Operation and drive control

In order to minimize experimental errors due to the manual positioning and movement of the substrate, an automated drive control program was implemented in a LabVIEW routine.

The software was designed to

- adjust coating speed and length,
- use a different speed for a second coating or a transfer into the drier,
- define a periodic movement inside the drier,
- return to the original position afterwards,
- and allow hands-free operation.

The experimental parameters are defined through a graphical user interface beforehand. The substrate is placed on the support plate and the slot-die system is vented and placed onto the substrate (Figure 2.6). With a tap on a foot pedal, the experiment is started and the substrate moves through the slot-die system into the drier. Here, the support plate moves back and forth as specified and returns to the original position afterwards.

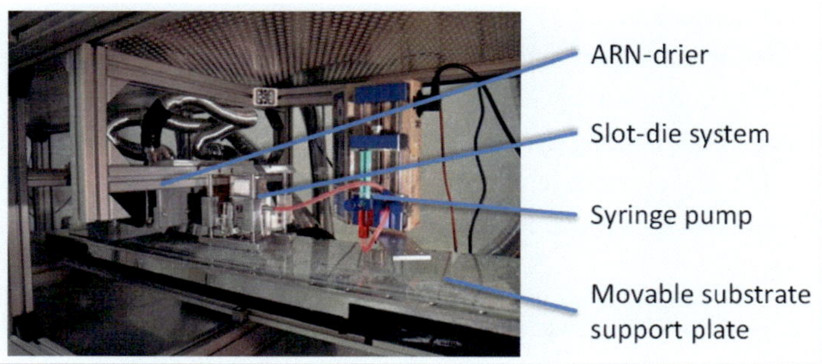

Figure 2.6: Photo of the Batch-Coaters coating chamber, illustrating its components.

2.3. Pilot coating and drying line (PC)

The pilot coating and drying line (PC), was developed in close cooperation by the plant manufacturers Coatema GmbH and TSE-Troller. The goal was the design of a set-up that can be adapted to the changing requirements in an experimental environment, while providing all elements required for technical coating and drying in a continuous R2R process.

2.3.1. Modular design

The coating line is structured in six modules. The drive module defines the speed, position, and tension of the web. The web treatment module cleans the substrate, using ionized air and a brush to remove electrostatically attached dust. Subsequently a corona unit is used to enhance the surface energy of the substrate. Currently, coating can be done with one of three different coating modules. The first coating module is a high precision premetered coating module for slot-die, slide, and curtain coating. This development coater (DC) is not connected to the frame of the main coating line, hence isolated from vibrations of the drive module. Alternatively it can be used for coating tests without substrate. The DC was provided by TSE-Troller within the scope of a research agreement. Coating module 2 and 3 are enclosed in a coating chamber which is permanently flushed with filtered air to remove solvents and dust particles from the atmosphere. Coating module 2 is a gravure / roll coating system designed by Coatema GmbH, which was subsequently customized for the requirements of PSC films (see Subsection 2.3.3). The speed and rotational direction of the coating roll can be controlled independently from the web speed or set as speed master[8]. The third coating module (Subsection 2.3.3) was designed to employ the slot-die coating system (Subsection 2.2.2) in a continuous R2R process. The drying module features two drier segments with a length of 1 m each and provides space for the implementation of additional drier segments.

---

[8]The "speed master" is the roll that sets the web speed of the plant.

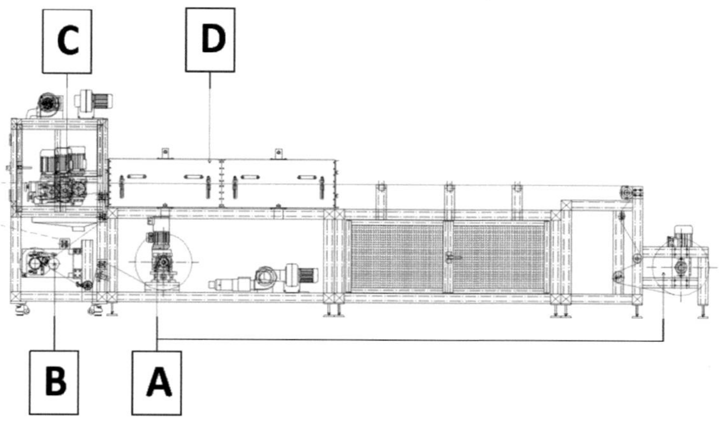

Figure 2.7: Technical drawing of the pilot-scale coating line, consisting of a drive module (A), a pre-treatment module (B), two coating modules (C) and a drier module (D).

The details of the basic construction and its individual parts are documented in Volume 1 and 2 of the pilot coater manual (Coatema-GmbH 2010), and the technical documentation of the development coater (Döll 2009). This work will only elaborate changes and customization of the original design and document the characterization of individual modules. Coating module 1 was used primarily for technical coating trials. All pilot-scale experiments described in this work were conducted in the enclosed coating chamber with module 2 or 3. A technical drawing and a photograph of this setup is shown in Figure 2.7 and Figure 2.8, respectively.

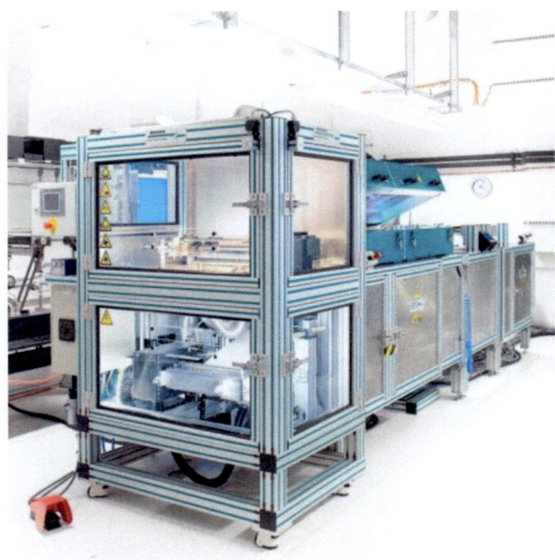

Figure 2.8: Photo of the equipment used in the pilot-scale experiments in this work.

To ensure defined conditions and reproducible results, all of these modules were characterized. The line speed of the drive module (Subsection 2.3.2), the gravure (Subsection 2.3.3), the slot-die coating module (Subsection 2.3.3), and the drier (Subsection 2.3.5) are discussed in the following subsections. The web treatment is closely related to the wetting properties of substrates and inks and is treated in Section 4.4 and Appendix 9.6.

2.3.2. Coating speed

The range of coating speed required in an experimental environment is significantly wider compared to a production environment. Low coating speeds are favorable for first experiments with new materials, where only few milliliters of ink are available. High coating speeds are necessary to test the stability limit of the process. To accommodate a wider speed range, servomotors with a velocity range from 0.2 m/min to 20 m/min were implemented. Coating speed determines coating thickness and process stability.

Figure 2.9 shows the relative deviation of coating speed from the set-point as a function of time for the standard configuration of the PC where the re-winder is set as speed master. Four set-point steps between 0.2 m/min and 1 m /min are shown here. Data for the full range is documented in Appendix 9.10.1. In this regime the coating speed shows an oscillation around the set-point value with a frequency of ~0.3 Hz.

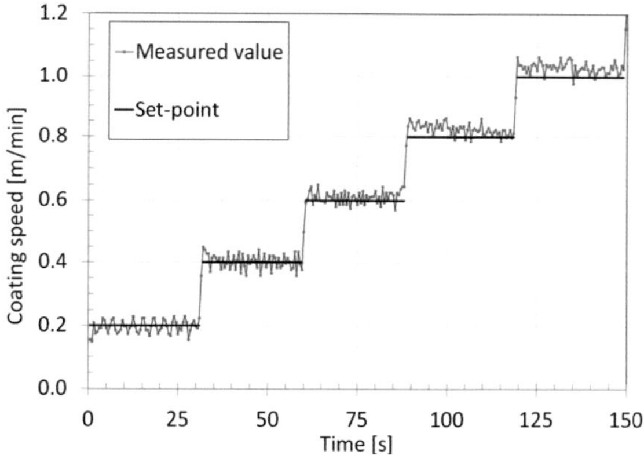

Figure 2.9: Set-point and measured value of the coating speed as a function of time for several steps in coating speed set-point. The re-winder was set as speed master. The sample rate was 500 ms, resulting in a resolution of 0.0064 m/min.

In Figure 2.10 the standard deviation is used as measure for the absolute deviation and the quotient of standard deviation and average value as a measure for the relative error. Though the absolute deviation increases, the relative error decreases with increasing speed and reaches a value of 1.2% at 3 m/min.

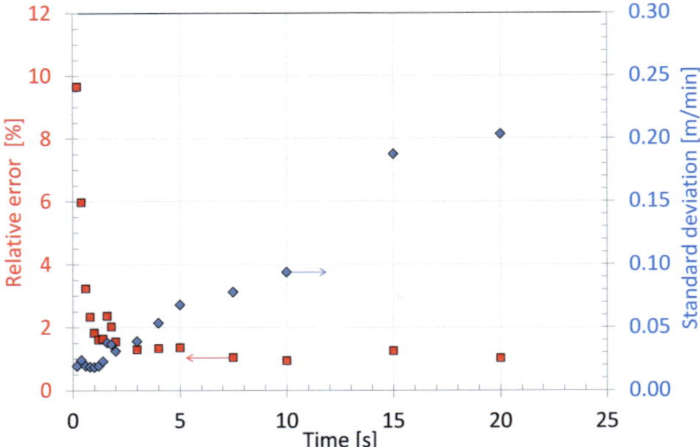

Figure 2.10: Absolute and relative deviation of the coating speed as a functions of coating speed. Though the absolute deviation increases, the relative error

The oscillation in speed is cause by the long distance (4 m) between the speed master (re-winder) and the coating module. Though the re-winder moves at a constant speed, the local web speed at the coating station may still oscillate due to the flexibility of the substrate. The oscillation of web speed below 3 m/min may result in cross web inhomogeneity during coating experiments as shown in Figure 2.11.

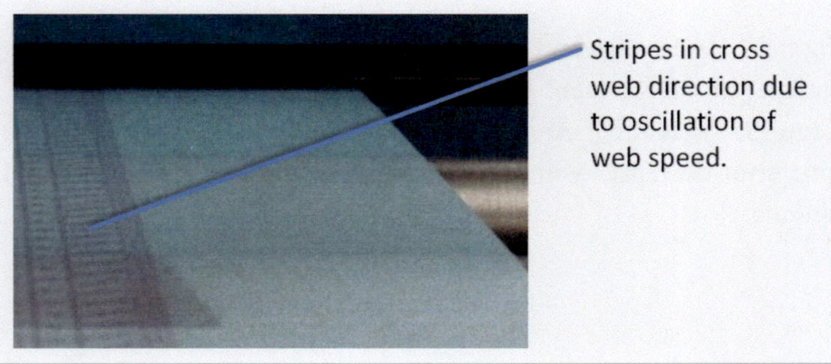

Figure 2.11: Stripes is cross web direction due to the oscillation of web speed when the speed is controlled by the re-winder. The photo shows slot-die coated polystyrene colored with Sudan-3.

To realize homogeneous coating speed at low velocities, the gravure coating roll was set as speed master, in all further experiments. The reduced distance between speed master and coating module (30 cm) resulted in improved coating speed homogeneity and no stripes in cross web direction were observed. The absolute speed of the coating line, using the gravure roll as speed master, was measured with a handheld tachometer (SMT-500C, Alluris). The measured value was found to be 20% higher than the set-point for all coating speeds (see Figure 9.38, Appendix 9.10.1). All coating speed values given in this work refer to the measured value.

2.3.3. Gravure coating system

During gravure coating, the ink is transferred from the reservoir into engraved depressions on the gravure rolls surface onto the substrate. Gravure coating takes place in the following four steps:

- Filling of cells,
- metering of the excess fluid with a knife,
- transfer of ink onto the substrate,
- and leveling of the ink on the substrate.

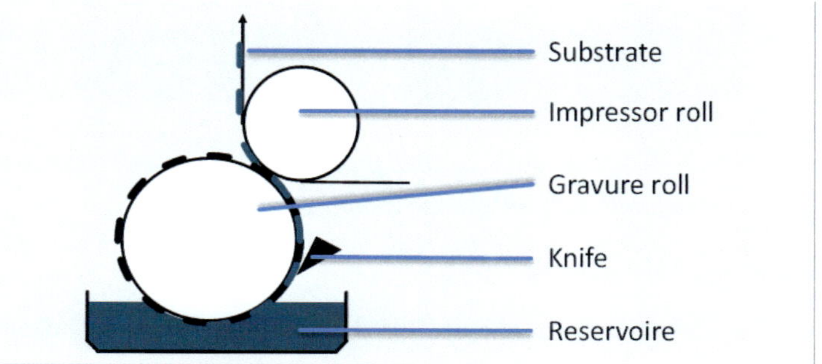

Figure 2.12: Schematic sketch of a gravure coating system.

A pneumatic adjustment of the force by which the gravure roll is pressed against the impressor roll was supplied with the original system. The customization of the gravure coating unit included: independent control of rotational velocity, pneumatic adjustment of the force applied to the metering knife, design of gravure pattern, and design of a closed reservoir with minimized hold-up.

Independent control of rotational velocity

The control panel for the gravure coating roll is shown in Figure 2.13. The optional independent control of the gravure roll (Figure 2.13 b) allows for a constant fixed rotational velocity (including negative velocities for reverse mode), fixed relative velocity, or fixed torsional moment. Though an independent control permits flexibility to all coating requirements, careless choice of process parameters (e.g. high relative speeds in combination with high normal forces between gravure and impressor roll) may lead to damage or destruction of the gravure cylinder or substrate. All experiments in this work were conducted with the gravure roll as speed master (Figure 2.13.a).

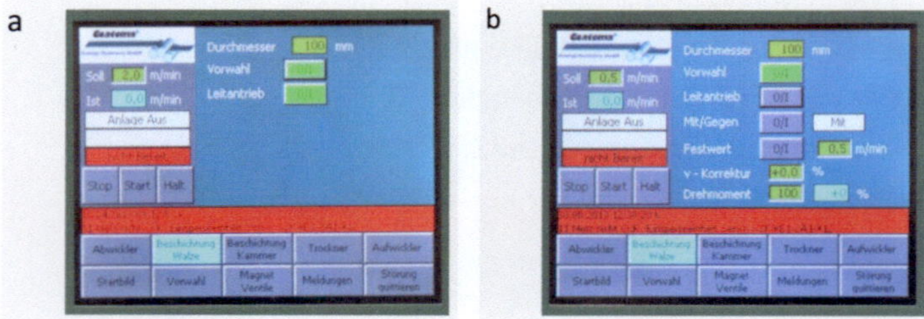

Figure 2.13: Screenshot of the control panel for the gravure coating roll as speed master (a) and with independent control setting (b).

Design of gravure pattern

Gravure cylinders can be manufactured by chemical etching, electromechanical engraving, or laser ablation. Highest homogeneity and abrasion resistance is achieved with laser engraved ceramic cylinders. Typical cell geometries for laser engraving are tri-helical cells[9], and closed or open hexagonal cells[10]. In open patterns, the adjacent cells are connected to each

---

[9] Linear grooves running parallel around the cylinder with a defined angle to the rotational axis.

[10] Spherical depressions which form a hexagonal pattern when overlapping.

34

other by channels, facilitating the leveling into a homogeneous film after ink deposition. The geometry and pattern of the engraved cells is one of the most important factors in gravure coating and is characterized by line spacing and specific cell volume. The specific cell volume is defined as volume of a cell divided by area of the cell projected onto the cylinder surface. Open hexagonal geometries are favorable for applications with high demands on homogeneity but are only available for a certain range of specific cell volumes. Specific cell volumes above 30 µm must be manufactured with tri-helical patterns and cell volumes below 10 µm require closed hexagonal patterns.

As a rule of thumb, the wet film thickness can be approximated as specific cell volume divided by two. Depending on which layer, the required wet film thickness in polymer solar cells may vary from ~2 µm (e.g. photoactive layers or charge adaptation layer) up to ~25 µm (e.g. electrode layers). In order to cover a range of wet film thicknesses, a gravure cylinder with five ribbons with different gravure pattern was manufactured by Zecher GmbH (Figure 2.14).

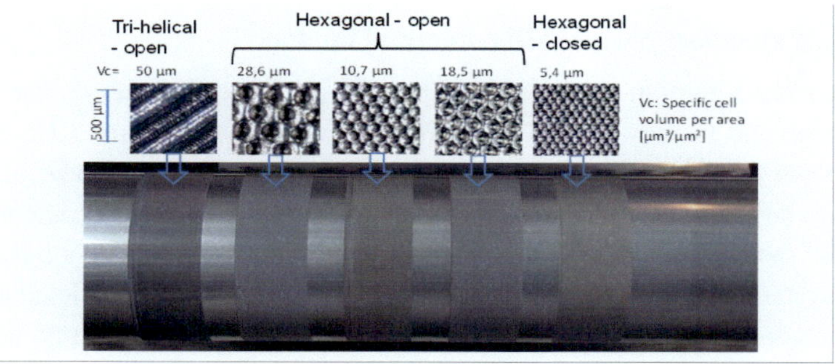

Figure 2.14: Microscope pictures of the cells on the gravure cylinder, designed for this work.

The geometric parameters of each ribbon are summarized in Table 2.2.

Using test inks and an open reservoir with a hold-up volume of ~500 ml, all gravure patterns can be tested in a single experiment. For experiments with material systems containing semi-conducting polymers, this system is not applicable due to the large hold-up volume and the solvent evaporation during the experiment.

Table 2.2: Overview of the geometric parameters of the gravure cylinder designed to realize wet film thicknesses from ~2.5- 25 µm.

Geometry	Specific cell volume V_c [µm]	Line spacing [L/cm]	Cell width A_1[µm]	Angle β [°]
Hexagonal closed	**5.4**	160	5.7	60
Hexagonal open	**10.7**	120	64.5	60
Hexagonal open	**18.5**	80	114.3	60
Hexagonal open	**28.6**	60	142.9	60
Tri-helical	**50.0**	30	192.9	45

Closed chamber system with minimized holdup

To decrease hold-up and solvent evaporation, a closed chamber system that supplies ink to only one of the ribbons was designed (Figure 2.15). The shape of the reservoir follows the curvature of the gravure cylinder to decrease hold-up volume further. A lateral guide-beam allows for positioning of the system under each gravure ribbon and micrometer calipers position the system in vertical direction defining the distance to the rotational axis. The micrometer calipers can be replaced by pneumatic cylinders to allow for a constant force adjustment instead of a defined distance. The fluid is contained within the reservoir to the sides by two Teflon seals, running on the smooth area between two ribbons. In coating directions, two 100 µm thick, stainless steel metering knifes control the film thickness and inhibit evaporation.

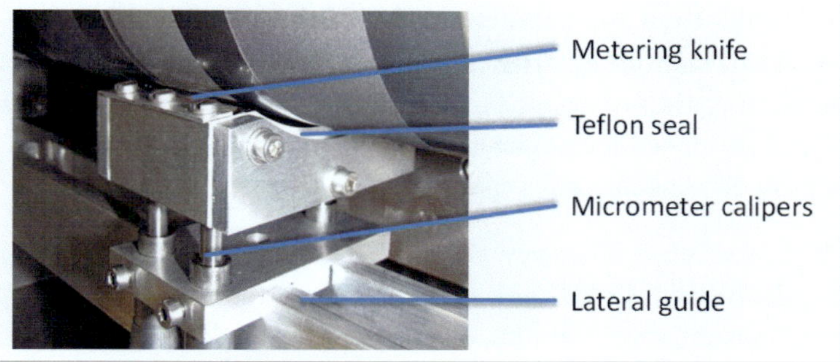

Figure 2.15: Photograph of the closed chamber system with a hold-up of 3 ml. The micrometer calipers can be replaced by pneumatic cylinders to operate at constant force rather than constant distance.

During gravure coating the configuration using pneumatic cylinders instead of micrometer calipers has delivered more reproducible results. The system can be used on a gravure free area of the cylinder for roll coating, as well. In this mode, film thickness is adjusted by a defined gap between cylinder and substrate rather than gravure geometry.

The implementation and test of the system was documented in a student project work by Peter Darmstädter (Darmstädter 2012) and first results will be discussed in Subsection 5.3.

2.3.4. Slot-die coating table

In order to use the slot-die system (see Subsection 2.2.2) in the pilot coating line, the PET substrate was led over a support table. The height of the table is adjusted to be as low as possible so that the web tension is sufficient to flatten out the substrate but the deflection of the substrate at the down-steam edge does not disturb the coating (~0.5 cm above normal substrate level). As the precision of the coating gap is influenced by the flatness of the table surface, an inlayed silicon wafer was used as support first. Later the silicon wafer was replaced by a more durable granite plate with a certified surface flatness of less than 0.4 µm.

2.3.5. Modification and characterization of the pilot line drier

The **pilot line slot-nozzle drier (PSN-drier)** consists of two segments with independent control of air temperature and volume flow rate. The nozzles consist of perforated metal plates with 4.65 mm wide slots in 90 mm distance from the substrate and a spacing of 300 mm.

The high particulate efficiency air filter (HEPA) reduces the volume flow of the inlet blower at 100% power to 143 m³/h at room temperature. To reduce the pressure drop, the filter cases were redesigned to allow for a utilization of the complete filter area (Figure 2.16, a). The nozzle exit velocity was increased further by reducing the nozzle length from 300 to 150 mm (Figure 2.16). In this configuration, the computed average heat transfer coefficient for an air temperature of 140°C reaches ~30 W/(m²·K) at 100% blower power (see Appendix 9.5.3).

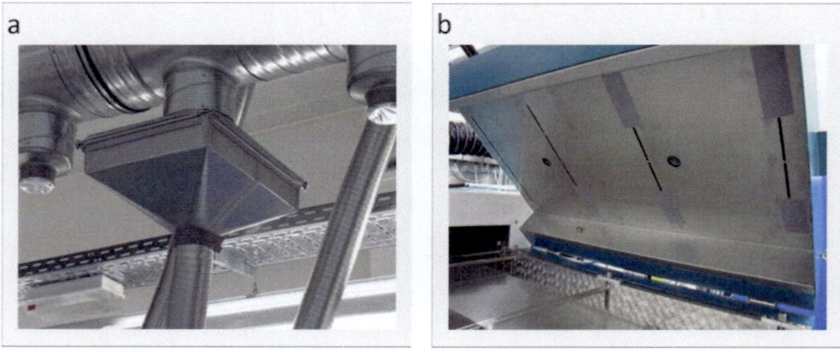

Figure 2.16: Redesigned HEPA filter case (a) and reduced slot-nozzle length (b).

Due to the different pressure drop over inlet and outlet blower, air is entrained from the environment reducing the actual air temperature in the drier. To minimize this effect, the outlet blower setting can be reduced to 60% power and the set-point temperature can be increased to attain the desired temperature. An asymmetry in nozzle exit velocity caused an inhomogeneous temperature profile within the drier. More air is entrained at the front inlet compared to the outlet of the drier. Similar average temperatures in both segments can be achieved if the set-point temperature of the first segment is 20°C above the set-point of the second segment. A "hot spot" at the center of the drier was reduced by closing the forth nozzle of the first segment. Calibration curves and characterization of heat transfer coefficients can be found in Appendix 9.5.3.

Though this basic design of the drier is not optimized with respect to energy efficiency, the high nozzle to web distance and relatively low nozzle exit velocity allow for homogeneous drying. The drier capacity was sufficient for most[11] experiments conducted in this work.

---

[11] For high speed coatings of water based polymer electrodes, the web had to be stopped after coating to increasing the residence time within the drier.

3. Experimental

The experimental details are equally important as the set-up for the reproducibility and scalability of the results. In this chapter, an overview over the material systems that were used in this work is given (Section 3.1). In Section 3.2 the experimental procedure for the preparation of inks, the application of films, and the pre- or post-treatment of films and substrates is outlined. Methods for characterization are presented in Section 3.3 with emphasis on methods that were established in the work group as part of this work.

3.1. Materials

The material systems used in this work can be structured according to their function within the solar cell into electrodes, photoactive layers, charge adaptation layers, or substrates. A tabular overview of coating inks is presented in Table 3.1. If the composition of the ink had to be adapted to an individual experiment, it is specified in the respective chapters.

Electrode layers were made from silver nanoparticles as reflective back electrodes, or from polymer nanoparticles as transparent conducting films. PEDOT:PSS applied as transparent polymeric electrode (PH500, PH1000), was received as water based dispersion. Small amounts of additives were used to improve its conductivity (Dimethylsulfoxide, DMSO) and wetting behavior (Triton X-100 (TX100)).

Photoactive layers were prepared from polymer-fullerene or polymer-nanoparticle blends. P3HT received as powder from BASF and Merck is labeled P3HT-P200 or P3HT-M, respectively. Solid PCBM (powder) was bought from Solenne BV. The photoactive polymers and fullerenes are typically dissolved in chlorinated solvents. Due to their toxicity, chlorinated solvents are undesirable in a technical process and non-chlorinated alternatives were studied as well. To adjust drying time and solubility, the following solvents with different boiling point were used: the chlorinated solvents Dichlorobenzene (DCB), Chlorobenzene (CB), and Chloroform (CF), and the non-chlorinated solvents Indane (IND), Xylene (XYL), and Toluene (TOL) (In order of decreasing boiling point). Inorganic semi-conducting nanoparticles (BayDots) were provided by Bayer Technology Services GmbH and stabilized with Pyridine (PYR).

Table 3.1: Overview of material systems used in this work.

Function (Type)	Name	Name (Specification)	Typical solvent
Reflective silver electrodes (Dispersion)	**PR-020**	Silver nanoparticles, InkTec	Isopropanol (IPA)
Transparent polymer electrode (Dispersion)	**PH500**	PEDOT:PSS, PH500, Clevios, Heraeus GmbH	Water, Dimethylsulfoxid (DMSO)
	PH1000	PEDOT:PSS, PH1000, Clevios, Heraeus GmbH	Water, Dimethylsulfoxid (DMSO)
Polymer : Fullerene (Solution)	**P200:PCBM**	P3HT, P200, BASF SE : [60]PCBM, Solenne BV	Chlorinated: Dichlorobenzene (DCB), Chlorobenzene (CB), Chloroform (CF); Non-chlorinated: Indane (IND), Xylene (XYL), Toluene (TOL)
	P3HT-M:PCBM	P3HT, Lisicon, Merck K.GaA : [60]PCBM, Solenne BV	
Polymer : Inorganic Nanoparticles (Hybrid/ Dispersion : Solution)	**P3HT-M:QD**	P3HT, Lisicon, Merck K.GaA : Quantum Dots, BayDots®, Bayer TS GmbH	Chlorobenzene (CB), Pyridine (PYR)
Electron blocking layer	**VPAI 40.83**	PEDOT:PSS, VPAI 40.83, Clevios, Heraeus GmbH	Water, Methanol (MeOH)
Hole blocking layer	**ZnO-alpha**	ZnO, diameter ~25nm, ZN2225, Nanoarc, Alfa Asear	Aceton(ACE), Propylene glycol monomethyl ether acetate (GPMEA), Unknown
	ZnO-gen	ZnO, diameter ~5nm, SAS Genes'Ink	Isopropanol (IPA), Unknown
Organic dummy system	**PS + Sudan3**	Polystyrene, Art.N.: 9151.1, Carl-Roth + Sudan3 (Art.N.: 4711.1, Carl-Roth)	Xylene (XYL)
Aqueaus dummy system	**PAA + Amidoschwarz**	Polyacrylamid, Art.N.:3048.1, Carl-Roth + Amidoschwarz, Art.N.: 9590.1, Carl-Roth	Water, Methanol (MeOH)

For **charge adaptation layers**, either hole conducting polymer particles or electron-conducting Zinc-oxide particles were applied. PEDOT:PSS, applied as hole-conducting layer, was diluted with demineralized water and the surface tension was adjusted with Methanol (MeOH). The hole blocking ZnO nanoparticles were received as stabilized dispersion from Alfa Aesar (ZnO-alfa) or Genes'Ink (ZnO-gen). ZnO-gen was diluted with Isopropanol (IPA), whereas ZnO-alfa phase separated with polar solvents and was diluted with Acetone (ACE).

Since material prices are high, **dummy systems** were used for basic coating experiments. Polystyrene (PS, Art.N.: 9151.1, Carl-Roth) colored with Sudan3 (1/30th of PS solid content, Art.N.: 4711.1, Carl-Roth) and Polyacrylamide (PAA, Art.N.:3048.1, Carl-Roth) colored with Amidoschwarz (Art.N.: 9590.1, Carl-Roth) were used as model systems for organic-solvent or water based inks, respectively.

All solvents were bought from Carl Roth or VWR with a purity of 99.5% or higher.

As **substrate**, float glass (120x60x1 mm³, Zitt Thoma GmbH) or Indium-tin-oxide coated glass (ITO-Glass; 12 Ω/sq, Vision Tec Systems) was used in the Batch-Coater. Continuous rolls of Polyethylene-terephthalate foil (PET) were used as substrate in the Pilot-Coater. From the large variety of available substrate foils, five samples were tested with respect to roughness (Ra values were determined with a confocal laser scanning microscope at the Institute of Production Science at KIT) and thermal shrinkage in machine direction (MD) and perpendicular to machine direction (TD) (Table 3.2). Hostaphan GN4600 was selected due to its acceptable roughness and shrinkage as well as good availability. As indicated in Table 3.2, Melinex O or Lumirror could be used as alternative.

For standard architecture cells, ITO-coated PET sheets (ITO-PET; 60 Ω/sq, 1ft x 1ft x 5 mil, Sigma-Aldrich) were glued or taped into the continuous PET carrier substrate.

Table 3.2: Test of substrate foils with respect to roughness and thermal shrinkage.

Sample	Price	Thickness	Roughness	15 min @150°C	
	[EUR/kg]	[μm]	Ra [nm]	MD [%]	TD [%]
Hostaphan GN4600	9.27	96	3.12	1.5	0.0
Melinex O	9.95	125	1.68	1.5	0.0
Lumirror 4001	8.20	96	1.70	1.0	0.0
Mylar A	8.80	100	4.08	1.0	0.5
PET GAG 340	6.45	100		4.0	1.0

3.2. Experimental procedure

Preparation of coating inks

The photoactive inks were prepared under nitrogen atmosphere and stirred for more than 12 hours at 40°C. Depending on the concentration and material system, even higher temperatures and longer stirring rates were applied to dissolve all agglomerates. A polymer to fullerene weight ratio of 1:0.8 was used in all experiments.

BayDot semi-conducting nanoparticles were received as concentrated dispersion in PYR. The particles were precipitated by adding 5 volume parts n-Hexane to 1 volume part BayDot (QD) dispersion and separated from the liquid by centrifugation for 25 min at 5000 rotations/min and subsequent decantation of the fluid. The particles were then dispersed in 90 v.-% CB and 10 v.-% PYR by shaking for 10 min on a vortex shaker and then immersed for 10 min in an ultrasonic bath. This dispersion was then mixed with a P3HT-M solution to yield a final concentration of 6 mg/ml P3HT-M, 54 mg/ml QD in 95 v.% CB and 5 v.% PYR. This concentration was used in all experiments with hybrid photoactive layers, unless stated otherwise.

All dispersions were submerged in an ultrasonic bath (S40-M, Elmasonic) for 10 min before experiments. To limit the temperature, a beaker with ice was inserted into the ultrasonic bath next to the samples. PEDOT:PSS dispersions were not filtered, because the filtering step contaminated the dispersion (observed by a decrease in surface tension) and resulted in an undefined decrease in solid content.

Coating procedure

The miniature slot-die coater (Figure 3.1, a) described in Subsection 2.2.2 was applied in a speed range of 0.5-20 m/min with gap widths of 25 µm, 38.1 µm or 50 µm. The gravure coating system (Figure 3.1, b; see Subsection 2.3.3 for details) was used in speed master configuration with the original blade adjustment at speeds between 1 and 5 m/min and a pneumatic pressure of 1 bar on the impressor roll. Knife coating was done with a commercial coating knife (ZAA2000, Zehntner) (Figure 3.1, c) at gap widths from 20 µm to 300 µm and a coating speed of 0.15 – 15 m/min (2.5 - 250 mm/s). If not specified otherwise, a volume of 100 µl was dosed with a microliter pipette and a gap of 100 µm was used. A simple paintbrush (Vega2000, Thayer & Chandler, Figure 3.1, d) was used at 1 bar nitrogen differential pressure for spray coating and a custom made spin coater at the Institute of Nanotechnology (KIT) (Figure 3.1, e) was employed at 500 1/s for PEDOT:PSS and 1000 1/s for active layers.

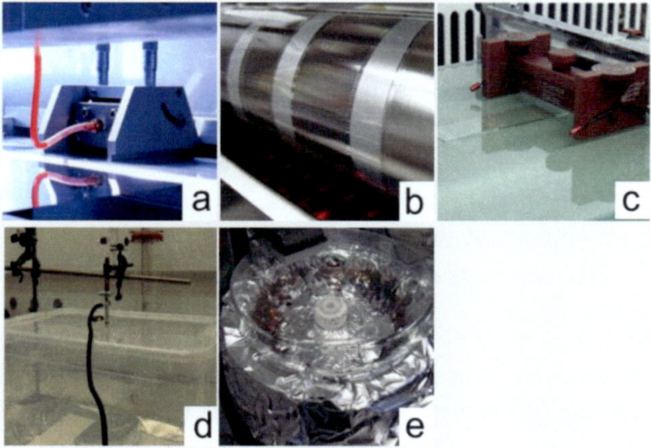

Figure 3.1: Experimental setup for slot-die (a), gravure (b), knife (c), spray (d), and spin (e) coating. The customization of slot-die and gravure coating is discussed in the previous chapter. The equipment for knife, spray and spin coating was not optimized.

Cleaning and plasma treatment

Glass substrates were exposed to ultrasound (S40-M, Elmasonic) for 10 min while immersed in Acetone and Isopropanol. For all water based coatings, substrates were treated for better wettability with an oxygen plasma generator (200 W, Pico-Diener) for 2 min at 80% power.

PET substrates were cleaned using an in-line cleaning tool (Typ P-Sh-N, Fa. Simco), removing electrostatic charges with an ion-spray-rod, and dust particles with a brush and exhaust ventilation. Corona (air-plasma) treatment was conducted with an in-line unit (P6000, 1,5 kW power supply Pillar; B10974-7, Narrow web treater station, Pillar) and an electrode distance of 1.09 mm.

Patterning of ITO coated Substrates

ITO coated substrates were patterned, in order to define the bottom electrode area and remove ITO at the position where the top electrode is connected. The areas where ITO should remain were covered with adhesive tape and the substrate was then immersed into concentrated hydrochloric acid (HCl-rauchend, 37% Rotipuran, Carl-Roth) for 120 s or 3 s for ITO-Glass or ITO-PET, respectively. The substrates were then rinsed thoroughly and the adhesive tape was removed. Prior to the standard cleaning procedure described in the previous paragraph, glue residue was removed. ITO-Glas was immersed into Toluene for 10 min and treated with ultrasound. ITO-PET was wetted with TOL, ACE and IPA and wiped carefully with cleanroom paper.

The device geometry for characterization at Bayer Technology Services required a circular ITO pattern. This was achieved by covering the ITO with circular adhesive paper sticker. Its chemical resistance was less than that of scotch tape, so that hydrochloric acid started to affect the conductivity of the ITO underneath.

Annealing procedure

All films were annealed under nitrogen atmosphere on a hot plate to remove residual solvents. Unless stated otherwise, electrode films were annealed for 30 min at 120°C and stored in Nitrogen atmosphere overnight prior to characterization.

3.3. Characterization methods

3.3.1. Characterization of wetting properties and surface tension

Static surface tension calculated from pendant-drop experiments and static contact angles from sessile-drop measurements were determined with a contact angle measuring instrument (DSA20S *EasyDrop, Krüss*; see Appendix 9.1 for details).

Characterization of substrates: Surface energy

The relationship of disperse and polar part of the surface energy (σ_S^D and σ_S^P), disperse and polar part of the surface tension (σ_L^D and σ_L^P), and contact angle (Θ) can be expressed by the Owens-Wendt-Rabel-Kälble (OWRK)-Modell ((Owens and Wendt 1969), see Appendix 9.6 for details):

$$\cos(\Theta) = 1 - \frac{2}{\sigma_L^P + \sigma_L^D}\left(\sqrt{\sigma_L^P \sigma_S^P} + \sqrt{\sigma_L^D \sigma_S^D}\right). \qquad (3.1)$$

Conventionally, the polar and disperse part of the surface energy are determined by a fit to a linearized form of Equation (3.1) (Hlawacek 2005). Here, the polar and disperse part (symbols in Figure 3.2) were varied to yield a minimal sum of square distance from the respective lines of the measured contact angle (lines in Figure 3.2). This method yields results with minimized error in σ_S^D and σ_S^P rather than minimized error of the linearized parameters. It was compared to the linear regression proposed by Hlawacek and led to similar results.

3. Experimental

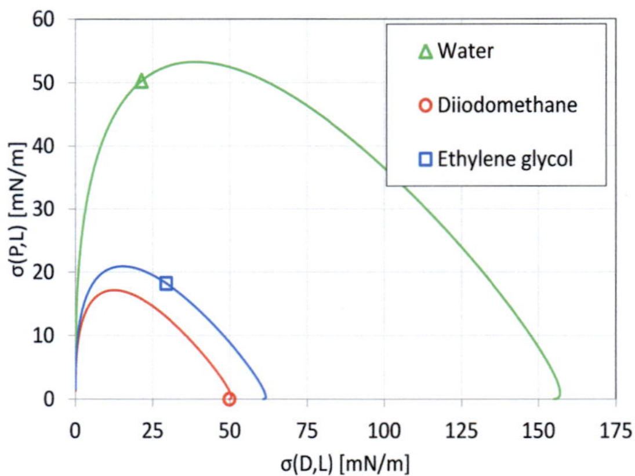

Figure 3.2: Lines predicted by the OWRK-model for the measured contact angles and polar and disperse part of the test liquids. The polar and disperse part of the surface energy were fitted by minimizing the sum of square distance between the points and respective lines.

The surface energy of substrates and films was determined by measuring the contact angle of two or three test fluids. Purified water (T172.1, Carl-Roth), Diiodomethane (1,8-Diiodomethane, 250295, Sigma-Aldrich), and Ethylene glycol (03750, Fluka) were applied when possible. As Ethylene glycol lies within the wetting envelope of many investigated surfaces, only water and Diiodomethane were used in this case. The characteristic data for the three test liquids is given in Table 3.3.

Table 3.3: Polar- and disperse parts (upper index P and D, respectively) of the surface tension of the test liquids Water, Diiodomethane, and Ethylene glycol.

Test Liquids	Water	Diiodo-methane	Ethylen gly-col
σ_L [mN/m]	71.6	49.8	47.5
σ_L^P [mN/m]	50.3	0	18.2
σ_L^D [mN/m]	21.3	49.8	29.3

Characterization of liquids: Surface tension

The surface tension of a liquid can be determined from a single pendant drop measurement (see Appendix 9.1). To determine the polar and disperse part of a liquids surface tension, however, at least two measurements are necessary. In this work one pendant drop measurement and two contact angle measurements on a Teflon block and a Silicon wafer were made to reduce experimental error. Curves predicted by the OWRK-Model for the two measured contact angles (blue and red line) and one pending drop measurement (green line) are shown in Figure 3.3. Ideally all lines should intersect in one point, determining the characteristic values of the investigated fluid. Due to experimental error, the three curves will not intersect in one point. Thus σ_L^P and σ_L^D were determined by minimizing the sum of square distance of the "point of measurement" to all curves.

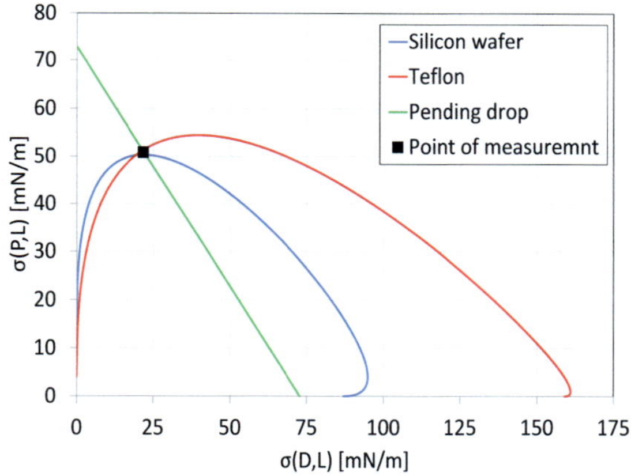

Figure 3.3: The curves of two contact angle measurements (Substrate 1 and Substrate 2) and the linear function given by a pendant drop measurement should ideally intersect at the characteristic point of the investigated fluid.

The characteristic data of the test substrates is listed in Table 3.4. The test substrates were cleaned by the same procedure outlined for glass and the Teflon substrate was additionally polished on cleanroom paper. The effects of cleaning procedure was investigated thoroughly in a student project work that was supervised by Peters (Griese 2011).

Table 3.4: Characteristic data of the test substrates: Silicon wafer and Teflon.

Test substrates	Wafer	Teflon
σ_S [mN/m]	39.45	18.43
σ_S^P [mN/m]	6.21	0.46
σ_S^D [mN/m]	33.24	17.97

3.3.2. Light absorption measurement

Light absorption was measured as a function of wavelength (300 nm - 800 nm; Genesys 10s, Thermo Scientific). Films on PET or glass substrate were positioned perpendicular to the beam.

The baseline absorption of PET and glass substrates was measured independently and subtracted from the absorption spectrum of the sample. As shown in Figure 3.4, the absorption of the substrate increases towards lower wavelength. A reasonable signal to noise ratio is achieved from 320 to 800 nm. Since the thickness and thus absorption of the substrates varies, a baseline corresponding to the substrates thickness was used. Substrate thickness was determined with a micrometer caliper (IP54, M&M, Messmittelonline or IP65, Quantum Mike, Mitutoya).

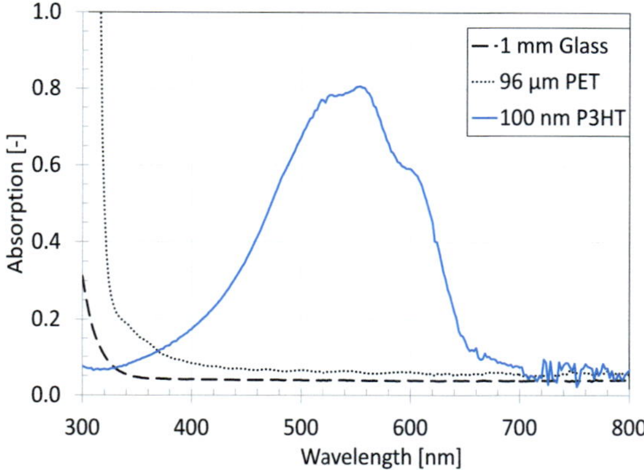

Figure 3.4: Comparison of the absolute absorption baseline spectra of PET and glass substrates to the spectra of a 100 nm P3HT-M film.

All spectra are averages of three or more samples to minimize experimental errors. Even though only films in a similar thickness range were compared, the absorption (A) was divided by dry film thickness (d) to correct for local film thickness variation, defining the specific absorption:

$$a = \frac{A}{d}. \tag{3.2}$$

Figure 3.5a shows the specific absorption of films of pure photoactive polymer (P3HT-M), fullerene (PCBM) and Quantum Dot nanoparticles (QD). The absorption of P3HT is characterized by a broad absorption above 400 nm and shoulders at 550 nm and 600 nm. PCBM-films are characterized by a peak at 350 nm, whereas QD films exhibit progressively increasing absorption towards lower wavelengths.

Figure 3.5b shows the specific absorption of blended films compared to a pure P3HT-M film. The characteristics of the individual components can be identified as superposition in the blend films.

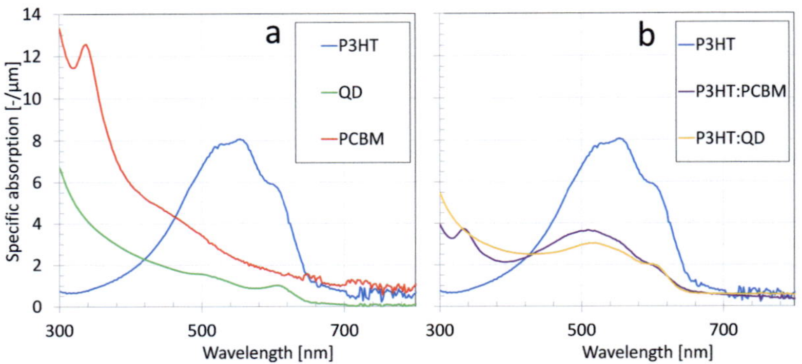

Figure 3.5: Specific absorption spectra of films of a: pure P3HT-M, pure Quantum Dots and pure PCBM, and b: blends of P3HT-M:PCBM and P3HT-M:QD.

3.3.3. Sheet resistance

A four-point resistance measuring setup was developed in collaboration with Baunach. The set-up is shown schematically in Figure 3.6.a. A voltage source (E3633A, Agilent) induces a current ($I_{1,4}$) through contact 1 and 4 into the sample. $I_{1,4}$ is calculated from the voltage drop U_R over as resistor R (10 kΩ) according to Ohm's law. The voltage drop within the film from contact 2 to 3, and the voltage drop over the resistor (U_R) are measured using a digital voltmeter (2700 Integra Series, Keithley).

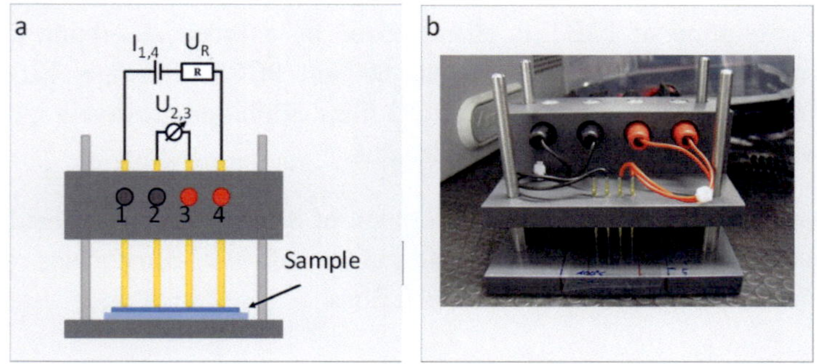

Figure 3.6: Schematic sketch (a) and photo (b) of the set-up for four point measurement of film conductivity.

For an infinitely large surface, the sheet resistance ($R_\square$) is defined by:

$$R_\square = \frac{\pi U_{12}}{I_{1,4}\ln(2)},\tag{3.3}$$

and the specific resistance ($SR_\square$) of a film with thickness d can be calculated by:

$$SR_\square = R_\square \cdot d.\tag{3.4}$$

Since the samples are not infinite in lateral dimension, the length b and width a of a sample were used for a geometrical correction as proposed by Smits (Smits 1958):

$$SR_\square = \frac{U_{2,3}}{I_{1,4}\cdot\frac{1}{\pi}\left[\frac{\pi\cdot s}{b}+\ln\left(1-e^{-4\cdot\pi\cdot s/b}\right)-\ln\left(1-e^{-2\cdot\pi\cdot s/b}\right)+\sum_{m=1}^{10}a_m\right]},\tag{3.5}$$

with

$$a_m = \frac{1}{m}e^{-2\cdot\pi\cdot\left(\frac{a}{s}-1\right)\cdot m\cdot\frac{s}{b}}\cdot\frac{\left(1-e^{-6\cdot\pi\cdot m\frac{s}{b}}\right)\cdot\left(1-e^{-2\cdot\pi\cdot m\frac{s}{b}}\right)}{\left(1+e^{-2\cdot\pi\cdot\frac{a}{b}m}\right)},\tag{3.6}$$

where s is the distance between the contact tips. The specific conductivity of a film (SC_{Film}) is the inverse of its specific resistance:

$$SC_{Film} = \frac{1}{SR_\square}.\tag{3.7}$$

As shown in Figure 3.6.b, the measuring head was implemented in a compact frame and connected with plug contacts. Hence, it can be transferred into a nitrogen glove-box easily. To measure conductivity under inert conditions, the system was operated using a wireless mouse inside the glove-box and a secondary monitor.

The set-up was successfully tested by comparing measured values of films with known conductivity (e.g. ITO-glas). A detailed characterization of the set-up was conducted in a diploma thesis supervised by Baunach (Ma 2012). Though specific resistance values calculated by Equation (9.4) and

(9.5) are independent of the lateral dimension, all values in this work were measured on 35 mm x 35 mm squares with a tip distance of s = 5 mm.

3.3.4. Film thickness

Film thickness was determined by measuring the step height over the edge of a scratch with a profiler (Dektak-3M, Veeco, or Dektak-XT, Bruker). The film was scratched down to the substrate with the dull edge of a spatula. A thin film of water and IPA (1:1 by volume) was used to fix the PET samples on a glass substrate. For plastic substrates, scratching the film may produce a systematic error if the substrate itself is scratched. It was thus necessary to test the validity of the method.

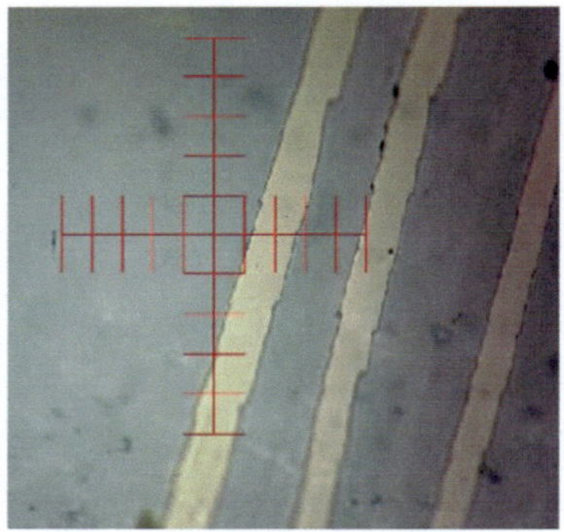

Figure 3.7: Top view of 3 scratches in a PEDOT:PSS film.

Film thicknesses of less than 1 µm on plastic substrate can be measured by light absorption. However, this technique is not applicable to photoactive films, as their light absorption may change as a function of process parameters, and does not work for transparent (PEDOT:PSS, ZnO) or opaque films (Ag-PR). Therefore, the organic dummy system (3 wt.-% PS and variable Sudan3 weight content) was used to compare film thickness from light absorption measurements to profiler measurements. Light absorption was averaged between 400 and 500 nm and calibrated using films on glass substrate.

Figure 3.8 compares film thickness determined by profiler and absorption measurements. A systematic error due to scratching of the PET surface would shift all points to higher values on the y-axis. The data points do not show a systematic deviation. Thus the profiler measurement can be used to determine film thickness on PET substrates and was applied for all measurements of opaque, transparent and photoactive films in this work.

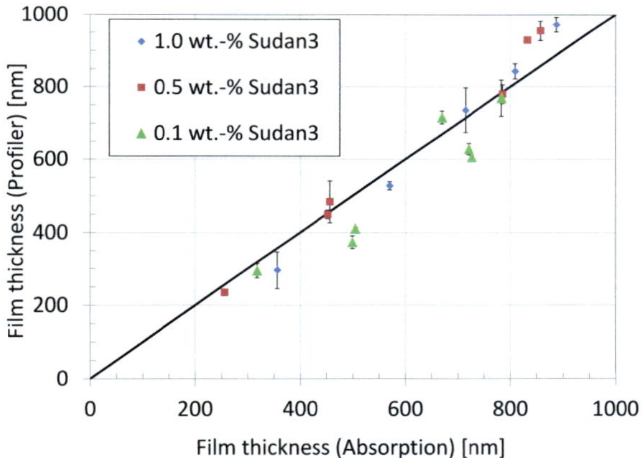

Figure 3.8: Comparison of film thickness of PS:Sudan3 films on PET determined by profiler measurement (Y-Axis) and light absorption (X-Axis).

3.3.5. Other characterization methods

Viscosity

Viscosity measurements were carried out with a cone plate rotational rheometer (*MCR101, Anton Paar GmbH*). Viscosity was measured at shear-rates from 0.1 to 10 000 1/s. Shear-rates below 100 /s were measured with a 60 mm cone diameter (CP-60-2/Q1, Anton Paar GmbH) and 30 s sample length. Shear rates above 100 /s were measured with a 25 mm cone diameter (CP-25-1, Anton Paar GmbH) and 10 s sample length.

Film topography measurement

Film topography was measured with an atomic force microscope (AFM; Dimension Icon, Veeco) in tapping mode. Roughness was determined from 30 x 30 µm scans, and 500 x 500 nm scans were used to visualize nano-scale topography. These measurements were conducted at the Institute for Microstructure Technology and supported by the Karlsruhe Nano and Micro Facility (KNMF) program.

Electric performance of solar cells

Due to the affiliation to different projects, experiments were conducted in collaboration with different institutions. Hybrid solar cells (P3HT-M:QD) were produced in standard architecture. Cathode evaporation and electrode deposition was done at the Bayer Technology Services Laboratory in Leverkusen, Germany. P3HT-M:OD and P3HT-M:PCBM cells for the comparison of coating methods were produced on ITO glass substrates, Aluminum electrodes were deposited at the institute of Nanotechnology (KIT), and cells were characterized at the Light Technology Institute (KIT). P3HT-P200:PCBM cells, manufactured on lab and pilot-scale, were processed with Calcium-Aluminum electrodes and characterized by Merck Chemicals Ltd. in Chilworth, Great Britain.

Though the individual measurement systems were different and cannot be disclosed. All measurements were done with calibrated or spectrally monitored set-ups with an AM1.5G solar spectrum under a radiation intensity of 1000 W/m². Cell area and additional information are provided with the individual results.

Preparation and handling of all films was conducted under ambient conditions, typical values of humidity and temperature are 30-60 % r.H. and 20°C-25°C, respectively. The samples were subsequently transferred into a nitrogen glovebox and annealed for 10 min at 120°C in inert environment to remove residual solvents. To ensure inert conditions during transportation, the samples were contained in a sealed plastic box (HPL816, Lock&Lock). Together with two oxygen/moisture absorber packs, the box was then placed in an aluminum packaging bag which was weld shut.

4. Material properties

In the last decade, many materials have been developed to enhance efficiency and lifetime of polymer based solar cells (PSCs) (see e.g. (Helgesen, et al. 2010)). These materials are typically prepared in small quantities and tested in laboratories, only. Therefore, properties needed for a scale up of the process are scarce if available at all. The most important properties for the coating process are viscosity and surface tension of the coating inks, and the surface energy of the substrate or underlying layers.

4.1. Introduction and state of the art

Ink viscosity and surface tension have been identified as important parameters that determine thickness (Chang, et al. 2005; Kopola, et al. 2010), process stability (Aernouts, et al. 2008; Voigt, et al. 2012), and morphology (Sun and Sirringhaus 2006; Hoth, et al. 2009b) of functional semiconducting films. However, the paper published in 2013 (Wengeler, et al. 2013) was the first systematic investigation of the influence of ink composition, and process conditions on the viscosity and surface tension of inks for polymer solar cells.

In this chapter viscosity data of typical ink compositions and a predictive model for commercial materials is presented (Section 4.2). Whereas high conductive polymer dispersions show a significant shear thinning effect, inks for photoactive, charge adaptation and electrode layers are assumed to be Newtonian fluids. Since solid content is typically less than 10 wt.-%, the viscosity of the solvent mixture constitutes a significant part of the overalls viscosity value.

Interfacial forces between solid, liquid and gas phase have a strong impact on the coating and wetting behavior. Whereas the surface energy of a solid can be assumed to be constant within the timespan of a typical coating process (Appendix 9.6.4), the surface tension of a liquid can be a dynamic function of the rate at which surface area is created and surfactant concentration (Hua and Rosen 1988). In general, the contact angle at a moving wetting line[12] is a function of rate of contact line movement, material properties, interface geometries and hydrodynamic fields in gas and liquid phase (De Gennes 1985). The detailed description of dynamic wetting behavior is subject of current research (see e.g. (Duvivier, et al. 2013; Snoeijer and Andreotti 2013)) and beyond the scope of this work. However, valuable insights can be gathered from determining static values of surface tension and contact angle. Surface tension of inks, dominated by solvent composition and surfactant concentration, is discussed in Section 4.3. Surface energy of substrates and activation of surfaces using corona treatment is addressed in Section 4.4.

Table 1 shows composition, density, viscosity, and surface tension for an exemplary system for each of the layers displayed in a polymer solar cell.

Due to the typically small solid content (for all systems except PR-020 and ZnO-alfa; see Table 4.1), the density (ρ) of the ink can be approximated as an ideal mixture of i solvents neglecting the density of the dissolved or dispersed solids:

$$\frac{1}{\rho} = \frac{x_{solid}}{\rho_{solid}} + \sum_i \frac{x_i}{\rho_i} \approx \sum_i \frac{x_i}{\rho_i} \quad \text{for} \quad x_{solid} \ll 1 \,, \tag{4.1}$$

where x_i is the mass fraction of solvent i. This approximation is often necessary since the exact density of the solid material is unknown.

In contrast to density, viscosity strongly depends on solid content and may be temperature and shear-rate dependent.

---

[12] A "wetting line" is defined as the dividing line between solid, liquid, and gas phase.

Table 4.1: Overview of typical values for fluid-dynamic properties of material systems applied in polymer based solar cells. See Table 3.1 for specification and function.

Name	Density (solid)	Solvent	Viscosity [mPas]		Surface tension	XSolid
	[g/cm³]		5 1/s	1000 1/s	[mN/m]	[wt.%]
PR-020	10.5	IPA	11.0	8.0	21.6	17.0
PH500	1.6	H2O/MeOH by volume: 89/11	60.3	12.5	69.2	1.3
PH1000	1.6	H2O/DMSO/TX100 by weight 94.5/5/0.5	24.9	11.5	30.0	1.8
P200:PCBM	1.3	CB	2.1	2.1	33.0	5.2
P3HT-M:PCBM	1.3	CB	4.4	4.3	33.0	1.3
P3HT-M:QD	4.0	CB	4.8	2.9	36.0	1.4
VPAI 40.83	1.6	H2O/MeOH by volume: 89/11	4.0	3.9	69.2	0.6
ZnO-alpha	5.6	ACE	5.0	4.4	19.2	40.3
ZnO-gen	5.6	IPA	13.2	11.9	22.0	1.4
PS + Sudan3	1.1	XYL	3.6	3.9	23.1	3.0
PAA + Amidoschwarz	0.9	H2O	70.2	24.5	71.8	0.3

4.2. Viscosity

Although many material systems applied in PSCs exhibit shear rate dependent behavior, it is most pronounced for high conductive polymer inks, such as PH500 or PH1000 (see Table 4.1). Low shear viscosity is relevant for the flow in tubes or the coating die's cavity as well as during the dewetting or leveling of films. In the die's coating slot and the coating gap (capillary gap between the die's lips and the substrate), shear rates of more than 1000 1/s are common. Viscosity may be five times higher for typical low shear conditions compared to high shear conditions, making a shear-rate dependent model inevitable.

4.2.1. Shear-rate dependent viscosity of PEDOT:PSS dispersions

The zero shear viscosity η_0 is determined by the type and solid mass fraction x_{Solid} (in wt.-%.) of the polymer in the dispersion and can be described at 20°C by an empirical fit to a second order polynomial:

$$\eta_{0;20°C}(x_{Solid}) = c_1 \cdot x_{Solid}^2 + c_2 \cdot x_{Solid} + \eta_\infty. \tag{4.2}$$

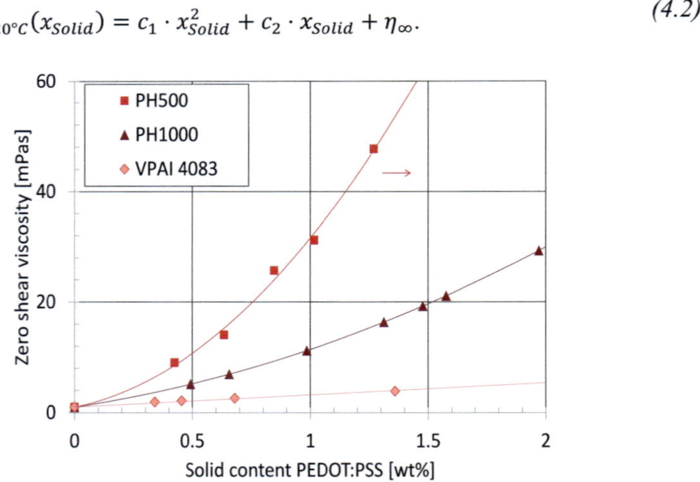

Figure 4.1: Plot of zero shear viscosity of various PEDOT:PSS types diluted with water. The data points originate from a least-sum-of-square-fit of the shear rate dependent viscosity (shown in Figure 4.2). The lines are second order polynomial fits (parameters given in Table 4.2).

The fit (lines) of Equation (4.2) to experimental data (symbols) for hree different PEDOT:PSS types is shown in Figure 4.1. The shear rate ($\dot\gamma$ in 1/s) dependence can be described by a Carreau-Yassuda model (see e.g. (Macosko and Larson 1994)) as a function of zero shear viscosity $\eta 0,20°C$, solvent viscosity $\eta\infty$, characteristic shear rate $1/\lambda_{Visc}$ and logarithmic slope (n-1):

$$\eta_{20°C}\left(x_{Solid}, \dot\gamma\right) = \eta_\infty + (\eta_{0;20°C} - \eta_\infty)\left[1 + (\lambda_{Visc} \cdot \dot\gamma)^2\right]^{\frac{n-1}{2}} . \qquad (4.3)$$

The temperature dependence can subsequently be super-positioned with an Arrhenius approach as

$$\eta(x_{Solid}, \dot\gamma, T) = \eta_{20°C}(x_{Solid}, \dot\gamma) \cdot e^{\frac{E_A}{R}\left(\frac{1}{T} - \frac{1}{293,14\,K}\right)}. \qquad (4.4)$$

The solid mass fraction of undiluted PEDOT:PSS dispersions was determined to be 1.97 wt.-%, 1.27 wt.-% or 1.36 wt.-% for undiluted PH1000, PH500 and VPAI4083, respectively. Model parameters for three commercial PEDOT:PSS dispersions are summarized in Table 4.2.

Table 4.2: Parameter values for concentration, shear rate, and temperature-dependent viscosity models. Parameters were fitted to data in the following range: $x_s<0.0197$, $10< \dot\gamma <10000$, and $10°C<T<50°C$.

PEDOT:PSS type	c_1	c_2	λ	n	E_A/R
	[mPas²/wt.-%²]	[mPas/wt.-%]	[s]	[-]	[K]
PH1000	4.064	6.351	0.0185	0.718	1205
PH500	22.24	8.308	0.054	0.688	1513
VPAI4083		2.1829	-	-	1906

Figure 4.2 shows experimental data (symbols) and the model (lines) for a PH500 dispersion at different temperatures and shear rates.

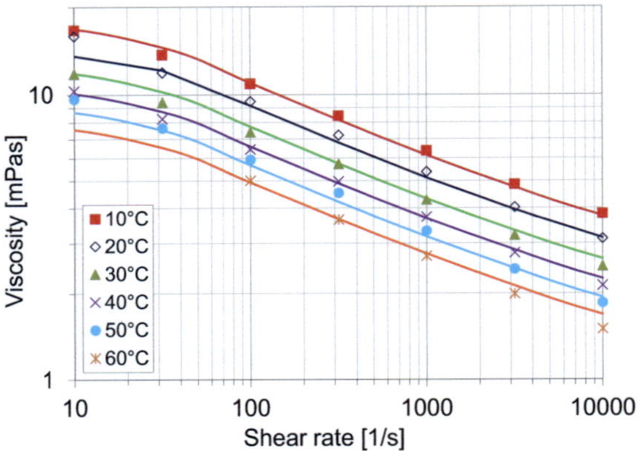

Figure 4.2: Viscosity data for PEDOT:PSS (Clevios PH500; x_{Solid} =0.564 wt.%) for different temperatures as a function of shear rate. The dispersion was diluted with water and methanol (11.1 vol.-% MeOH). Measurements (symbols) were carried out with a cone plate rotational rheometer (MCR101, Anton Paar GmbH) and fitted by a Carreau-Yasuda-Modell (lines).

The shear thinning behavior is caused by shear induced structural changes, a phenomenon that is often observed in dispersions (Willenbacher 2013). Polymer fullerene solutions and all other dispersions show no or only insignificant shear thinning behavior within the investigated range of solid contents.

4.2.2. Newtonian viscosity of solvent based inks

The photoactive polymer fullerene blends, dissolved in organic solvents, exhibit Newtonian behavior at low solid concentrations and their viscosity can be described at 20°C by:

$$\eta_{20°C}(x_{Ful}, x_{Solid}) = \eta_\infty + k_1 \cdot x_{Solid} + k_2 \cdot x_{Solid}^2 + f_1 \cdot x_{Ful} , \tag{4.5}$$

where x_{Solid} and x_{Ful} are the solid mass fraction (in wt.-%) of polymer and fullerene, respectively. The model parameters for an exemplary material system of P3HT-P200 and PCBM are given in Table 3. Note that for each new polymer (including polymers of similar structure but different molecular weight) the parameters k_1 and k_2, and for new fullerene derivatives the parameter f_1, have to be measured.

Table 4.3 also gives parameters for ZnO (Zn2225) and Silver nanoparticle (PR-020) dispersions and Polystyrene as a model system.

Table 4.3: Parameter values (Equation 5) for concentration dependent viscosity of solvent based inks for OPV at 20°C. Viscosity was fitted for a solid mass fraction $0<x<x_{S,Max}$ and may be considered as shear-rate independent in this concentration range.

Material	k_3 [mPas²/(wt.-%)²]	k_2 [mPas/wt.-%]	f_1 [mPas/wt.-%]	$x_{S,Max}$ [wt.-%]
P3HT-P200, BASF	0.8287	-0.0629		2
PC$_{60}$BM, Solenne B.V.			0.454	4
ZnO, Alfa Asear	0.0029	-0.0175		40
PR-020	0.0218	0.0555		20
Polystyrene	0.1723	0.4196		8

Based on viscosity data of P3HT-P200 in CB (blue diamonds) and PCBM in DCB (red squares), the model parameters were fitted (solid blue and red line). With these parameters and Equation 5 the viscosity of a blend, with mass ratio 1:0.8 of P3HT-P200:PCBM in CB, can be estimated. The model (broken green line) is in good agreement with the measured data (green triangles).

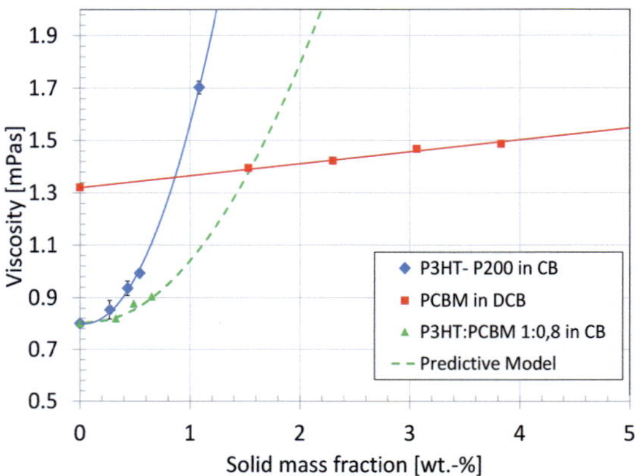

Figure 4.3: Viscosity of photoactive materials, P3HT-P200 (blue), PCBM (red) and P3HT-P200:PCBM (green, dissolved in CB or DCB. The additive model for the blend (broken green line) is in good agreement with measured data.

Once the parameters are determined for a given polymer, it is possible to estimate the viscosity at various compositions (polymer fullerene ratio, solid mass fraction and solvent composition) without further measurements as demonstrated in Figure 4.3.

4.2.3. Solvent viscosity

The infinite shear viscosity η_∞ in Equation (4.2) and (4.5) is assumed to be equal to the solvent viscosity and can - depending on the composition - significantly affect the overall viscosity value. Data for relevant binary solvent-water-mixtures (taken from literature *(Gonzalez, et al. 2007) and own measurements) are shown in Figure 4.4. In PEDOT:PSS dispersions DMSO and EG are often used as conductivity enhancing additives (see e.g. (Dimitriev, et al. 2009)) and alcohols are often added to improve wetting behavior (see e.g. (Voigt, et al. 2012)).

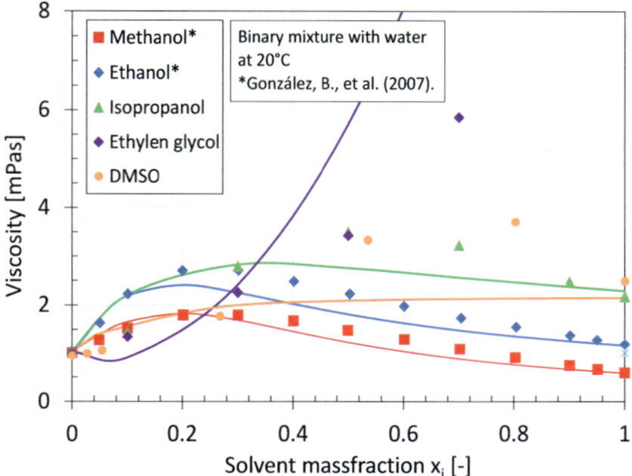

Figure 4.4: Experimental data for the viscosity of binary solvent-water mixtures. The data for Methanol and Ethanol was taken from Gonzales et. al. (Gonzalez, et al. 2007), solid lines represent binary fits based on the model of Teja and Rice (Teja and Rice 1981).

Whereas the addition of small amounts (< 5 wt.-%) of high viscous additives, Ethylene glycol, DMSO and Triton X (TX100 not shown here) has little impact on overall viscosity, the addition of 20 wt.-% Ethanol can increase the solvent viscosity by a factor of 2.7 (an increase of 1.7 mPas absolute).

The viscosity of pure solvents can be computed as a function of temperature using the correlation and data compiled by Yaws (Yaws 1999). Teja and Rice (Teja and Rice 1981) employ a generalized corresponding states method to predict viscosity of liquid mixtures (solid lines in Figure 4.4) based on a binary interaction coefficient $\Psi_{i,j}$ (see Table 4.4). Whereas most solvent mixtures can be predicted with reasonable accuracy, highly non-ideal mixtures (such as EG:H2O) may have to be fitted to higher order polynomial equations.

4.3. Surface tension of coating inks

Figure 4.5 shows surface tension of PCBM solutions (diamonds) and VPAI dispersions with different solid content. Surface tension of both inks varies by less than 3%. For typical solid concentrations the impact of solid content on surface tension is insignificant compared to the impact of solvent composition.

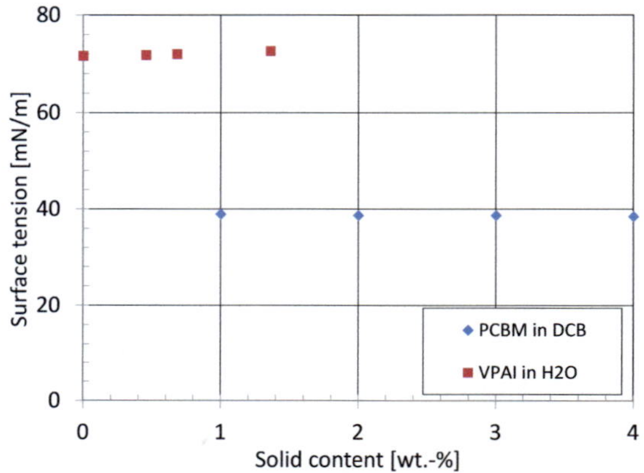

Figure 4.5: Compared to the impact of solvent composition, the change in surface tension due to the addition of PCBM and PEDOT:PSS is negligible for typical solid contents.

The temperature dependence of pure solvents is tabulated and fitted to a linear equation by Jasper (Jasper 1972) and a model for the surface tension of binary mixtures with water is given by Conors and Wright (Connors and Wright 1989):

$$\sigma = \sigma_{H_2O} - \left[1 + \frac{e_1 \tilde{x}_{H_2O}}{1 - e_2 \tilde{x}_{1H_2O}}\right] \tilde{x}_{Solvent}(\sigma_{H_2O} - \sigma_{solvent}). \tag{4.6}$$

Here $\tilde{x}_i$ and σ_i are the molar fraction and surface tension of component i, and e_1 and e_2 are binary interaction parameters (see Table 4.4).

Table 4.4: Interaction parameters for the calculation of viscosity and surface tension of binary solvent-water mixtures. [1]Teja and Rice (Teja and Rice 1981), [2]Connors and Wright (Connors and Wright 1989), [3]this work.

	Viscosity	Surface tension		
	$\Psi_{i,j}$	σ_2 [mN/m]	e_1	e_2
Methanol	1.34	23.5 [1]	0.899 [2]	0.777 [2]
Ethanol	1.36	23.4 [1]	0.963 [2]	0.897 [2]
Iso-Propanol	1.27	22.4 [3]	0.984 [2]	0.970 [2]
Dimethylsulfoxid	1.08	43.8 [3]	0.869 [3]	0.603 [3]
Ethylenglykol	0.82	45.9 [3]	0.826 [3]	0.867 [3]

Figure 4.6 shows the predicted surface tension of water solvent mixtures as a function of solvent mass fraction. The surface tension shows a progressive increase towards a solvent content of zero for all mixtures.

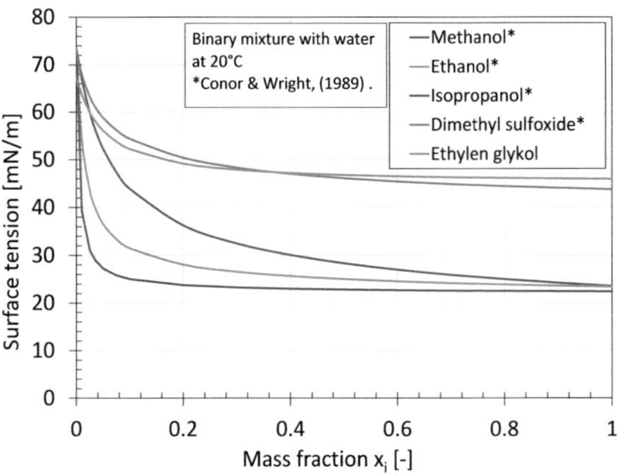

Figure 4.6: Surface tension of binary solvent:water mixtures. Taken from literature(Connors and Wright 1989) and own measurements.*

In contrast to the low surface-tensions (SFT) of alcohols (< 25 mN/m), the SFT of DMSO and Ethylenglycol (EG) mixtures remain above 40 mN/m. In an inverted device architecture, a high conductive PEDOT:PSS dispersion with 5v.-% DMSO as an additive has to be coated on top of the BHJ layer. To enable wetting on low surface-energy (SFE) substrates, such as an untreated P3HT-P200:PCBM photoactive layer ($SFE_{P3HT:PCBM,Untreated}$ = 29.7 mN/m), the SFT of these dispersions has to be decreased by adding a surfactant. Figure 4.7a shows the SFT of an aqueous dispersion with 90 vol.-% PH1000 and 5 vol.-% DMSO as a function of the mass fraction of the non-ionic surfactant TritonX100 (blue diamonds). Compared to TX100 mixtures with pure water (red squares), the critical micelle concentration is shifted to a higher TX100 mass fraction of 0.1 wt.-% (= 0.17 mol/l). Up to a solid mass fraction of 0.5 wt.-% TX100, the specific conductivity of coated films does not decrease with surfactant concentration (Figure 4.7.b). However, higher concentrations lead to a decrease in specific conductivity and should be avoided in a technical process.

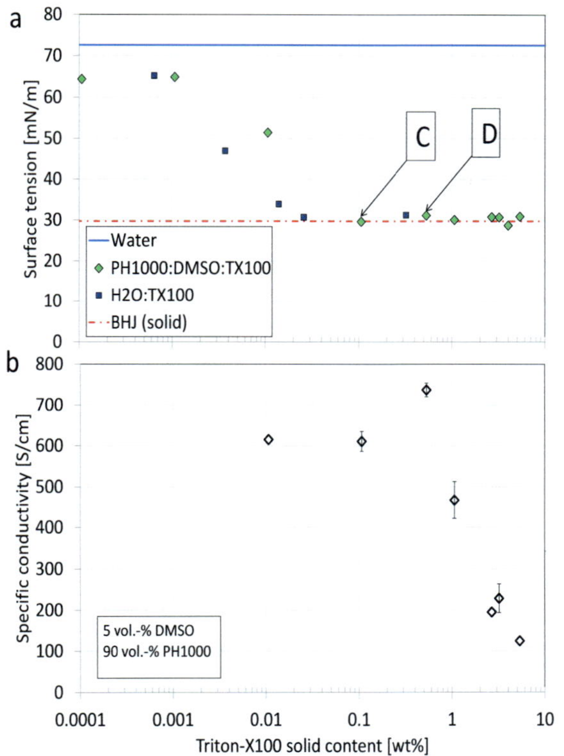

Figure 4.7: Surface tension (blue diamonds; a) and specific conductivity of the coated film (b) for a dispersion of high conductive polymer (90 vol.-% PH1000, 5 vol.-% DMSO) as a function of TX100 solid mass fraction. The surface tension of a water TX100 mixture (red squares; a), pure water (solid line) and the surface energy of a P3HT-P200:PCBM BHJ (dot-dash line) are shown as reference.

The surface tension of PH1000:DMSO:TX100 is equal to that of the BHJ layer at a TX100 concentration of 0.1 wt.-% (Figure 4.7: point-C). However, the film clearly de-wets on the BHJ layer at point-C (Figure 4.8.c). Good wetting is achieved for a coating speed of 5 m/min at a concentration of 0.5 wt.-% TX100 (Figure 4.7 point-D; Figure 4.8.D).

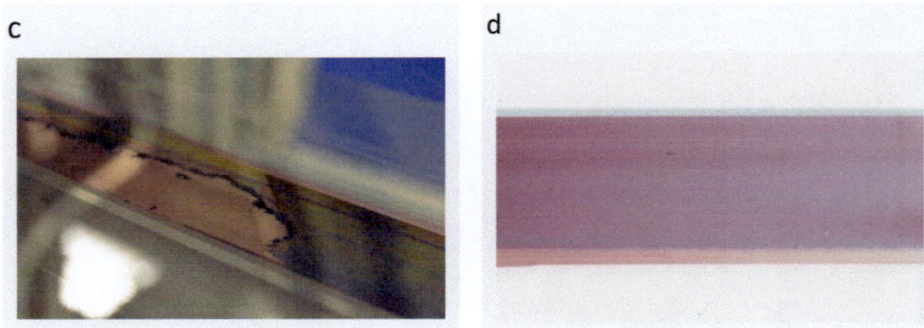

Figure 4.8: PH1000:DMSO:TX100 coatings with 0.1 wt.-% (c) and 0.5 wt.-% TX100 on a P3HT-P200:PCBM layer.

In solutions with surface active additives, such as TX100, the dynamic surface tension may be significantly higher compared to the static surface tension. This is due to the finite time in which surfactant molecules diffuse to the surface and depends on the rate at which surface is created (Eastoe and Dalton 2000). However, a surfactant content of 0.5 wt.-% TX100 led to sufficient reduction in dynamic surface tension while sustaining the high specific conductivity of the film at a coating speed of 5 m/min.

At higher coating speed, or lower wet film thickness the rate at which surface area is created increases and 0.5 wt.-% TX100 may not be enough to ensure good wetting. Higher TX100 contents are undesired because they will lead to a reduction in specific conductivity (Figure 4.7.b). Possible solutions may be the selection of a different additive, or an activation of the surface of the substrate or BHJ-film.

4.4. Surface energy of substrates and functional layers

Voigt et al. (Voigt, et al. 2012) showed that a treatment of the photoactive layer with oxygen plasma may improve wetting without sacrificing device efficiency. In continuous coating lines an air-plasma treatment (corona treatment) is often used to modify surface energy. Here, the residence time is a function of the coating speed so that the power of the corona generator has to be adjusted to achieve a desired surface activation. Wolf and Sparavigna (Wolf and Sparavigna 2010) expressed the intensity of corona treatment as product of power times residence-time divided by area. In this paper we chose instead to define the treatment intensity by:

$$PD_{Corona} = \frac{P \cdot \tau^3}{A_{Corona} \cdot \sigma_{Solid,Min}}, \qquad (4.7)$$

with P being the power of the corona generator, τ the residence time, A_{Corona} the area on which the corona is applied, and σ_{Min} the minimal surface tension (here equivalent to the untreated value). The residence time is a function of web speed (u) for a given length in web direction (L) and width perpendicular to the web direction (B) of the corona gap. Equation (4.7) becomes:

$$PD_{Corona} = \frac{P \cdot L^2}{u^3 \cdot B \cdot \sigma_{Solid,Min}}, \text{ with } \tau = \frac{L}{u} \qquad \text{and} \qquad A_{Corona} = L \cdot B. \qquad (4.8)$$

Figure 4.9 shows the surface energy and its disperse and polar part of corona treated P3HT-P200:PCBM blends as a function of treatment intensity. The overall surface energy can be increased up to a value of 42.5 mN/m and the steepest increase occurs at low treatment intensities ($PD_{Corona} <$ 200 000 s²).

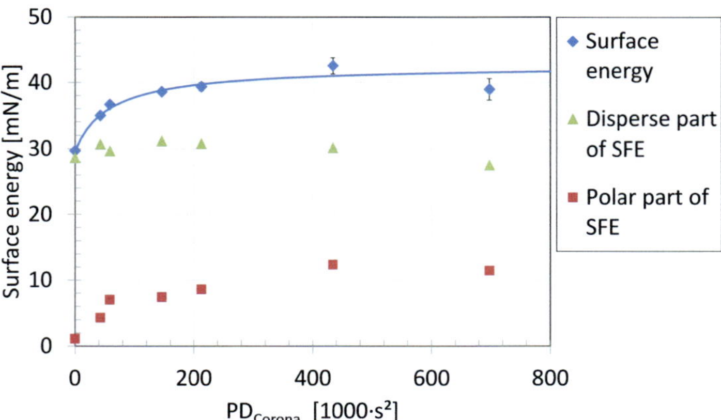

Figure 4.9: Surface energy of a plasma treated P3HT-P200:PCBM film as a function of treatment intensity (blue diamonds). The solid line is a fit to Equation (4.10). Disperse-part and polar-part of the SFE are shown by triangles and squares, respectively.

The surface energy (SFE; σ_{Solid}) can be expressed in a dimensionless form as:

$$\Pi_{SFE} = \frac{\sigma_{Solid} - \sigma_{Solid,Min}}{\sigma_{Solid,Max} - \sigma_{Solid,Min}} .$$ (4.9)

The effect of corona treatment of PET (Melinex OD) substrates with variable coating speed (triangles) and variable power settings (squares) is shown in Figure 4.9. The same correlation (blue line in Figure 4.9 and Figure 4.10) can be used to express the increase in surface energy for P3HT:PCBM and PET:

$$\Pi_{SFE} = \frac{PD_{Corona}}{\frac{1}{f} + PD_{Corona}},$$ (4.10)

where f is the initial slope. f depends on the experimental set-up (e.g. efficiency of corona generator) and the properties of the treated substrate (e.g. thickness, material). Using this correlation it is now possible to select the required corona power for a given coating speed in order to achieve a desired increase in SFE.

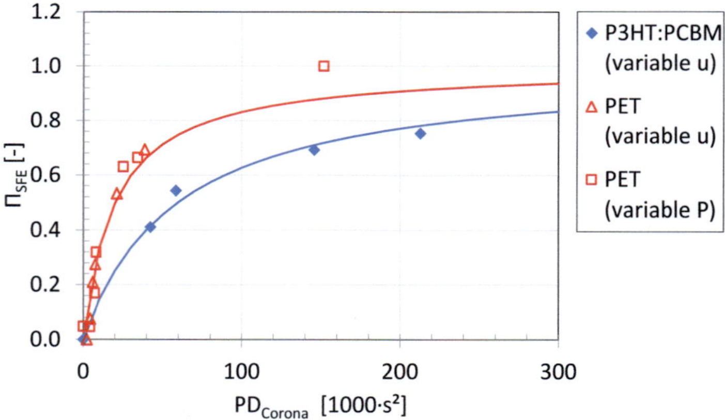

Figure 4.10: Dimensionless surface energy (defined in Equation (4.9)) as a function of treatment intensity (defined by Equation (4.8)). Diamonds represent the surface energy of a photoactive blend (P3HT-P200:PCBM) whereas triangles and squares represent a PET substrate treated with var-

iable speed (triangles) and power(squares), respectively. The solid lines represent least sum of square fits of Equation (4.10) using f as a fit parameter.

The correlation is valid for power intensities <200 000 1/s² and in-line corona treatment systems where the coating takes place mere minutes after the corona treatment. For plasma ovens, which are often used to treat glass substrates in laboratories, the surface tension decreases linearly over time after the treatment (see Appendix 9.6.4).

The half-life time ($t_{1/2}$) of the plasma treatment lies in the order of minutes to days, depending on the type of substrate and the ambient humidity during processing (e.g. ITO $\sim t_{1/2}$=90 min at $\sim$75 % rH).

Based on the models presented in this section, viscosity and surface tension of coating inks can be predicted as a function of solid content, shear rate (for PEDOT:PSS), temperature and solvent composition. These values were used in the following chapters to design coating die cavities (Appendix 9.3.1), to present process parameters in a dimensionless form (Chapter 5), and to discuss the stability of films during drying (Section 6.4). The manipulation of an ink's surface tension with surfactants and a substrate's surface energy by plasma treatment was shown exemplarily for the wetting of high conductive PEDOT:PSS dispersions on PET and BHJ layers.

5. Coating of functional films for polymer solar cells

The material properties, determined in the previous chapter, in combination with the film thickness and homogeneity requirements, determined by device physics, differ significantly from conventional coating processes. In addition to film homogeneity and defect density, the functional properties of the film, such as conductivity, and specific light absorption become important for the determination of the optimal coating method and the best operating point.

5.1. Introduction and state of the art

In conventional coatings, solid content is typically higher than 10 wt.-% resulting in a viscosity of more than 20 mPas. Romero, et al. (Romero, et al. 2004) discusses slot-die coating of "dilute solutions" with solid content ranging from 20.0 to 30.1 wt.-%. In contrast, for solutions of photoactive polymer fullerene blends, the solid content is limited by solubility to approximately 3 wt.-% at room temperature, resulting in viscosities of less than 10 mPas (see Section 4.2). The precise solubility limit depends on solvent and temperature and is often not specified. Estimates of solubility limits in various solvents were published for P3HT:PCBM by Machui, et al. (Machui, et al. 2012). Higher temperatures may increase solubility but may also increase degradation and complicate handling by requiring temperature controlled reservoir, tubes, and coating tools.

At constant surface tension and coating speed, a lower viscosity results in a lower capillary number. The capillary number is defined as a ratio of viscous to capillary forces and will be small for typical line speeds (v=5-20 m/min) when coating the hole-conductive or active layers of polymer based solar cells:

$$Ca = \frac{\eta u}{\sigma}; \ Ca << 0.1 , \tag{5.1}$$

where η is the viscosity, u the coating speed, and σ the surface tension. At low capillary numbers the coating flow is dominated by surface tension forces. Lee, et al. (Lee, et al. 1992) showed that at low capillary numbers (<100) the minimal critical wet film thickness in slot-die coating becomes a function of the capillary number. Although he used test fluids with viscosities as low as 10 mPas to reach Capillary numbers down to 0.05, the absolute wet film thicknesses of 200 µm is more than 10 times higher compared to film thicknesses expected in polymer solar cell production.

It is thus important to investigate the impact of viscosities below 10 mPas and film thickness on process stability and film formation. In Section 5.3, coating methods for the hole-conductive and active layer in polymer and hybrid solar cells are compared, using an identical material batch and - as far as possible - identical process conditions.

The semi-conducting polymer and hybrid films were characterized with respect to homogeneity and surface roughness using a profiler and an atomic force microscope (AFM). As well, light absorption spectra are shown as a function of coating method, giving a first indication whether a specific coating method has an influence on opto-electric film properties (Section 5.4).

In Section 5.5 the impact of viscous shear on film morphology and opto-electric properties is investigated for a slot-die coating process. Typical inks, consisting of polymer solutions or nanoparticle dispersions, are coated at variable coating conditions, exerting a different amount of viscous shear on the coating fluid. The performance of the functional films is discussed with respect to the demands of conducting, semi-conducting, and photoactive films and their implication for a large-scale production process of PSCs.

5.2. Specific experimental details

Since the work in this chapter is affiliated to different research projects, different material systems were used in Section 5.3 - 5.4 and 5.5. General information on experimental set-up, procedure and characterization can be found in Chapter 2 and 3. Specific details for the individual sections are provided below.

Section 5.3 and 5.4

In Section 5.3 and 5.4, P3HT-M, QD and PCBM were used for the active layer. Cells were produced in standard architecture and VPAI40.83 was used as hole conducting layer. The thickness of the hole conducting layer (HCL) was set to ~30 nm and that of the hybrid (P3HT-M:QD) photoactive layer (AL) to 80 nm, for all coating methods except spray coating. Spray coated films were thicker, but due to their extreme roughness a precise determination of their film thickness was not possible. To allow for a direct comparison of organic and hybrid photoactive films, the P3HT-M content of the inks and the coating parameters were held constant, resulting in a dry film thickness of 40 nm for the organic (P3HT-M:PCBM) material system. The knife coated films were 20 nm thicker than expected, because the calibration was based on the model system. The average thickness of knife coated films for the organic and hybrid photoactive films was 47 nm and 99 nm, respectively. Though the coatings processed in this study were large area (4 - 60 cm²) the final fabricated solar cells had an active area of only 0.24 cm² due to constrains of the experimental setup for cathode evaporation (100 nm Aluminum) and cell characterization under AM1.5 radiation (spectrally monitored ORIEL solar simulator). All layers were annealed for 1 hour at 110°C under nitrogen atmosphere after coating.

Section 5.5

In Section 5.5 a bulk hetero-junction of P3HT-P200 and PCBM was used. The photoactive material was dissolved in chlorinated solvents (Chlorobenzene (CB), and Chloroform (CF)).

The transparent conductive films were produced from Poly-3,4-ethylenedioxythiophene:poly-styrenesulfonate (PEDOT:PSS; PH1000, Heraeus) which was received as a water based dispersion. Dimethylsulfoxid (DMSO) was added to improve conductivity and either Methanol (MeOH), or Triton X-100 (TX100) was added to improve wetting behavior.

Reflective electrodes were coated from silver nanoparticle dispersion (PR020, InkTec). The nanoparticles were received as an Isopropanol (IPA) based dispersion and diluted further with IPA to adjust material properties and solid content.

All films were dried with an average heat transfer coefficient of ~30 W/(m²·K) in the PSN-drier (Appendix 9.5.3) at 140°C (set-point) air temperature, unless specified otherwise.

Electrode films were characterized, before and after an annealing step (30 min @ 120°C in Nitrogen atmosphere). Even after the annealing step, desorption of residual water resulted in a regressive increase of specific conductivity during the next 24h. If not specified otherwise, the annealed samples were thus kept under nitrogen atmosphere for more than 24 h prior to characterization. The photoactive films were annealed (30 min at 120°C) and characterized under ambient conditions (~30-70 % rH).

5.3. Evaluation of coating methods

A variety of coating methods are, in principle, suitable for preparation of one or more layers in solar cells, as discussed by Krebs (Krebs 2009a). Here we chose to focus on knife coating due to its simplicity and low hold-up, slot-die coating due to its superior homogeneity and controllability, and spray coating because it allows coating of curved or rough substrates. Spin coated samples were prepared as a reference to laboratory scale experiments.

5.3.1. Spin coating

The majority of reports on solar cells still employ spin coating as method for film deposition. Spin coating is a well-established process in the semi-conducting industry and it is easily set-up in laboratories or even in a glove box. It is however limited to small rigid substrates and not compatible with a continuous production process. There are three major problems associated with the transfer to a continuous process:

- De-wetting may occur in technical processes, while it is suppressed by fast drying rates and leveling forces during spin coating.
- All process parameters are interdependent in spin coating so that the impact of individual parameters in a technical process cannot be determined.

- Process values are a function of the position on the substrate and may lead to inhomogeneous distribution of material properties on the substrate.

Figure 5.1 shows a schematic cross sectional sketch of a differential volume element of a spin coating film. The centrifugal force per unit volume $\vec{g}$ depends on density of the fluid (ρ), the spin frequency (Ω) and the distance from the rotational axis (R):

$$\vec{g} = \rho\Omega^2 R. \tag{5.2}$$

Expressed in polar coordinates, the lubrication approximation (Kistler and Schweizer 1997) yields:

$$\eta\frac{d^2u}{dy^2} = -\rho\Omega^2 R. \tag{5.3}$$

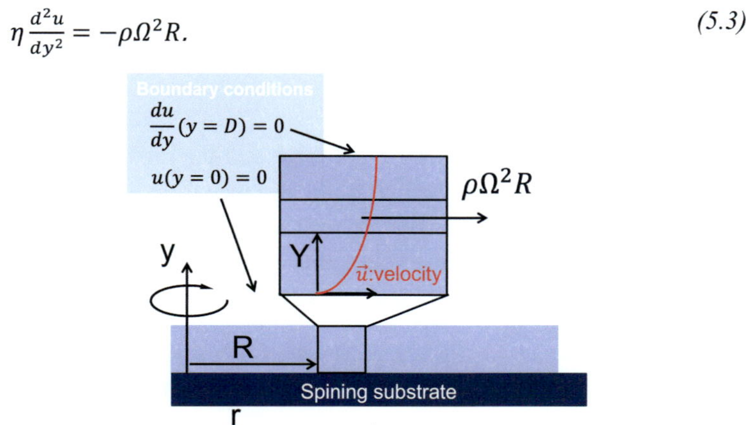

Figure 5.1: Schematic sketch of a differential element of a film during spin coating.

Assuming zero shear-rate at the fluid air interface (y=D; see Figure 5.1), integration in y direction defines the shear rate as a function of position in y direction (Y), radius, and angular velocity:

$$\frac{du}{dy} = \frac{\rho\Omega^2 R}{\eta}(D - Y). \tag{5.4}$$

Further integration (wall adhesion at the substrate (y=0)) yield a function for the radial velocity at position Y:

$$u = \frac{\rho \Omega^2 R}{\eta}\left(D \cdot Y - \frac{Y^2}{2}\right),$$

(5.5)

and the rate of volume flowing from the element

$$Q = \int_0^D u\, dy = \rho\frac{\Omega^2 R D^3}{3\eta},$$

(5.6)

where D is the height of the film.

Since Q scales with D³, the volume flow is highest where the film is thickest. Thus, the centrifugal force acts as driving force for the liquid distribution and leveling of the film. As substrates are typically rotated until dry, this leveling force remains active until viscosity is high enough to inhibit unwanted flow (e.g. Marangoni flow, coffee stain effect, de-wetting), resulting in high process stability.

On a spinning substrate the centrifugal force and the relative velocity between air and substrate are functions of the distance from its center. During coating, a convective flow transports material from the center of the substrate toward the edges. Process values thus change as a function of radius and a differential fluid element passes through different zones depending on all process parameters.

As a free-flow coating method, there is only one process parameter apart from material properties that can be used to influence the process. For given material properties and ambient conditions, spin frequency determines both coating flow and drying rate. It is thus not possible to investigate the influence of one individual physical parameter such as drying rate without changing other parameters as well.

5.3.2. Knife coating

Knife coating can be done with a small amount of ink (~100 µl/ 25cm² sample) on laboratory equipment. In contrast to spin coating it is compatible with a continuous roll-to-roll process and thus very attractive for technoligies during scale-up. A sketch of a knife coating process is given in Figure 5.2.

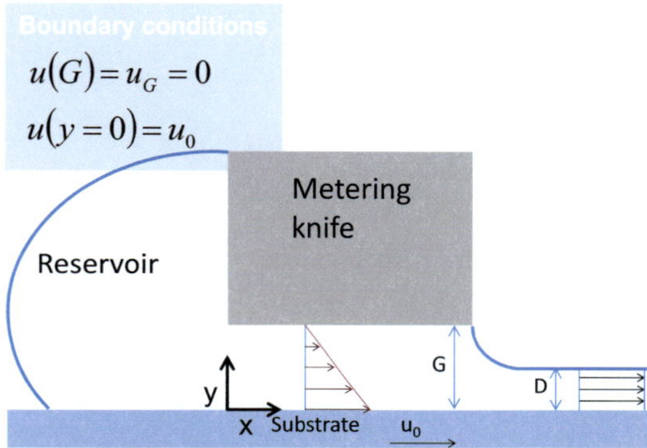

Figure 5.2: Schematic cross section sketch (Not to scale) of a knife coating process. Velocity profiles for a fully developed flow are indicated by black arrows.

In case of a fully developed coating flow, the lubrication theory predicts a Couette-flow (linear velocity profile between the knife and substrate boundary), yielding a wet film thickness of half the coating gap (indicated in Figure 5.3 by horizontal lines):

$$D = \frac{G}{2} \qquad (5.7)$$

Figure 5.3 shows the wet film thickness (D) as a function of coating velocity for knife-coated films determined at three different knife gaps (G) for the dummy system for the active layer (Model-AL). Similar curves were observed for PEDOT:PSS, P3HT-M:PCBM, and P3HT-M:QD inks (see Appendix 9.10.3).

For high coating speeds (>200 mm/s or >12 m/min; Ca >0.06) the experimentally determined thickness approaches the theoretically predicted value (solid lines for D=G/2). However, for lower knife coating speed, the metering effect of the knife coating gap G is diminished and the wet film thickness decreases.

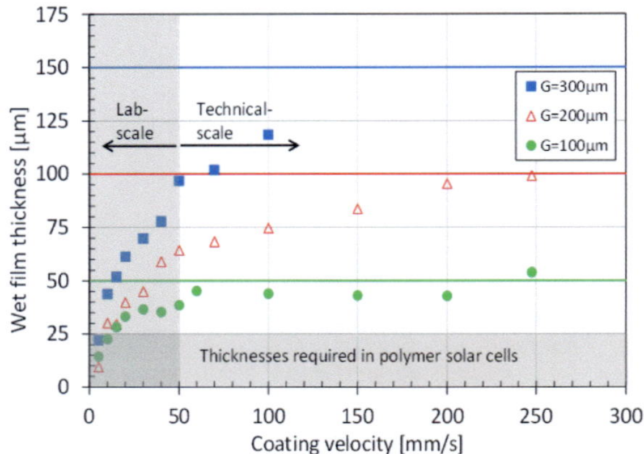

Figure 5.3: Wet film thickness as a function of coating velocity for knife coating of 3 wt-% polystyrene in xylene with three different gap widths. 100 µl solution was deposited manually as reservoir to coat 25x50 mm float-glass on a laboratory knife coater (Figure 3.1.C).

At low coating velocities, coating thicknesses required for polymer based solar cells (<25µm), can thus be produced with coating gaps much larger than twice the wet film thickness. On the one hand, as the gap width is no longer the most important metering parameter, the film homogeneity is less dependent on mechanical precision of the coating knife. On the other hand, the resulting film thickness depends on coating speed, feed system, surface tension, and viscosity of the applied ink as well as the surface energy and geometry of the knife. Therefore film thickness cannot be predicted for a new material system or coating knife and has to be calibrated for each new material system.

To reach higher coating velocities, as may be desired in a technical process, the gap width has to be in the order of two times the desired wet film thickness. As films of less than 5 µm are common, this is challenging with respect to mechanical precision. Small variations in gap width due to substrate inhomogeneity, knife elasticity, angle, or surface topography will have a large impact on wet film thickness.

5.3.3. Spray coating

In the simple spray coating setup used here, the volume flow of coating ink was adjusted by flow-rate and pressure of the dispersing air in the spray nozzle. An unknown fraction of ink was then deposited on the fixed substrate, resulting in a decreasing film thickness from the center of the sprayed area to the sides. Here, samples were taken from a 20 by 20 mm² square in the center of the sprayed area, where film thickness was constant within the limit of experimental uncertainty. Film thickness was then adjusted varying the deposition time. Figure 5.4 shows the surface profiles of spray coated hole conducting layers (PEDOT:PSS) with thickness of 115 nm (solid black line) and 270 nm (dotted black line) measured by a 3M Dektak (Veeco) profiler.

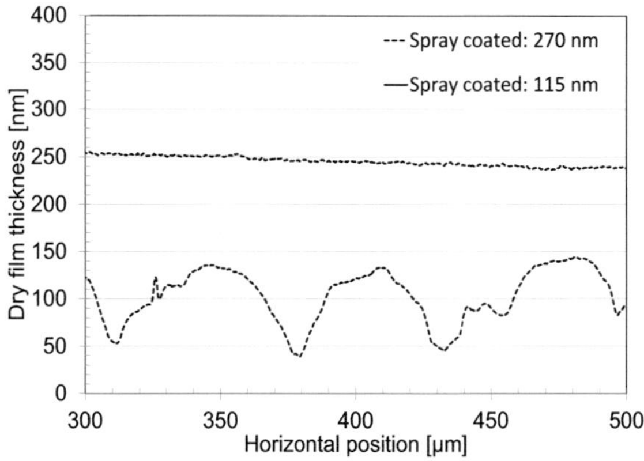

Figure 5.4: Cross profile measurements of spray coated PEDOT:PSS films with different thickness (Dektak 3M, Veeco).

The 115 nm spray coated layer shows variations in the same order of magnitude as the film thickness, whereas the 270 nm spray coated film shows a relative deviation of only ~4%. The same tendency was observed for spray coated P3HT-M:PCBM and P3HT-M:QD active layers (see Appendix 9.10.6). Apparently, there is a minimal spray coating thickness for which leveling takes place. This critical leveling thickness as well as the magnitude of the deviation is probably a function of the experimental setup and could be optimized by spray coating with saturated gas, resulting in better leveling, and more energy input during atomization (e.g. ultrasonic spray system nozzle diameter), resulting in smaller droplet diameters.

5.3.4. Gravure coating

Gravure coating can be realized by a variety of process configuration, depending on the number of rolls, rotational velocity, and size of the transfer rolls. In this work, the most basic configuration with one 100 mm transfer cylinder, moving at the same velocity as the web, was used (see Subsection 2.3.3).

The wet film thickness D can be expressed as

$$D = \frac{\Phi}{s} \cdot V_C,$$ (5.8)

where Φ is an empiric transfer coefficient, s the speed ration of web to roll ($s=u_{Web}/u_{Roll}$) and V_C the specific cell volume ($V_C = V_{Cell}/A_{cell}$) (Kapur 2003). For direct gravure coating without differential speed, s=1 and Φ typically assumes values around 0.5 (Lightfood 2009).

The transfer coefficient is a function of capillary number, linear force on the coating knife, cell geometry and coating velocity (Kapur 2003; Lightfood 2009). At low capillary numbers, and high linear forces Φ has a constant value which is determined for a specific material system by cell geometry and coating velocity (Lightfood 2009). The wet (left axis) and dry (right axis) film thickness of PEDOT:PSS films is shown for different cell geometries as a function of their specific cell volume in Figure 5.5. The solid line represents Equation (5.8) with $\Phi = 0.5$ and s=1 as a reference. The transfer coefficient decreases with increasing specific cell volume but the absolute wet film thickness still covers a range from 4 to 11 µm.

The error bars in Figure 5.5 show the standard deviation of more than 5 film thickness measurements and represent a measure for local film homogeneity. For this material system, the lowest film homogeneity was observed for the hexagonal open gravure with V_C=28.6 µm and a wet film thickness of 6.5 µm. The homogeneity is determined by the leveling and wetting behavior of the film.

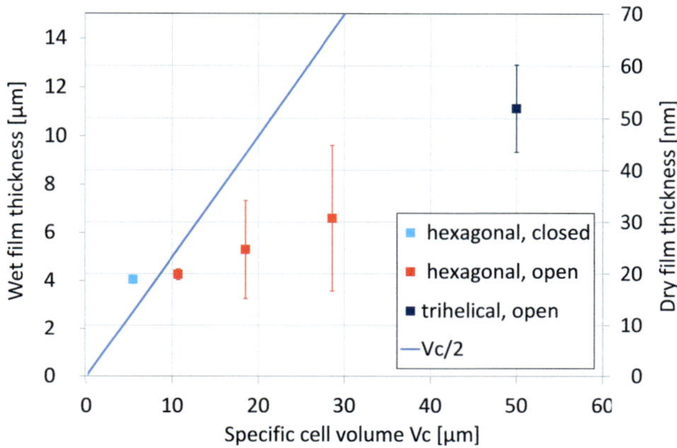

Figure 5.5: Wet and dry film thickness of PEDOT:PSS (VPAI40.83) for different gravure ribbons as a function of their critical cell volume. The layers were coated from a water based dispersion with Water:MeOH ratio of 8:1 by volume and solid content of 0.6 wt.-%.

The dry film thickness of 19 nm to 52 nm represents realistic thicknesses for hole-conducting layers in PSCs. A thickness of 30 nm, recommended in most standard architecture cells, can be produced with a specific cell volume of 28 µm. Similar dry film thicknesses were obtained by Kopola, et al. (Kopola, et al. 2009) who used specific cell volumes from 35 µm to 48 µm for gravure coated PEDOT:PSS layers.

For the organic dummy system, no homogeneous films could be prepared with the tri-helical gravure pattern. The open and closed hexagonal gravures showed similar results as the PEDOT:PSS films and produced dry film thicknesses between 60 nm and 200 nm. Film thickness as a function of specific cell volume is provided in Appendix 0.

Though homogeneous films could be produced (Figure 5.6.A), low film thicknesses often caused de-wetting and insufficient leveling (Figure 5.6.B) higher film thicknesses led to a better leveling, but also to smearing to the sides (Figure 5.6.C).

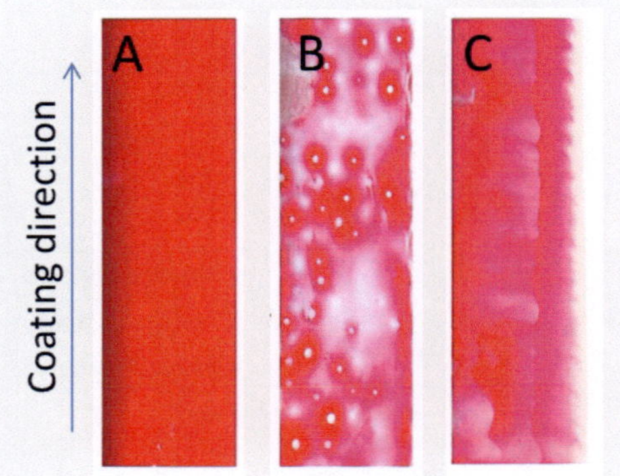

Figure 5.6: Scans of different films of the organic dummy system produced by gravure coating. Film thickness, surface energy and drying kinetic determine whether the film shows good leveling and wetting.

In summary, film thicknesses suitable for the preparation of hole-conducting and photoactive layers were produced by gravure coating and specific cell volume can be varied to adjust the film thickness. The homogeneity achieved with the gravure system was poor and requires further optimization. It remains questionable, whether methods that apply an inhomogeneous film in form of droplets or cell-pixels and rely on leveling can compete with processes that apply a homogeneous wet film.

5.3.5. Slot-die coating

In slot-die coating, the volume flow is adjusted with a metering pump, and thus directly defines the wet film thickness for a given coating width and line speed. The minimal achievable wet film thickness is limited by the stability of the liquid bridge between the die lip and the moving substrate (see Figure 5.7).

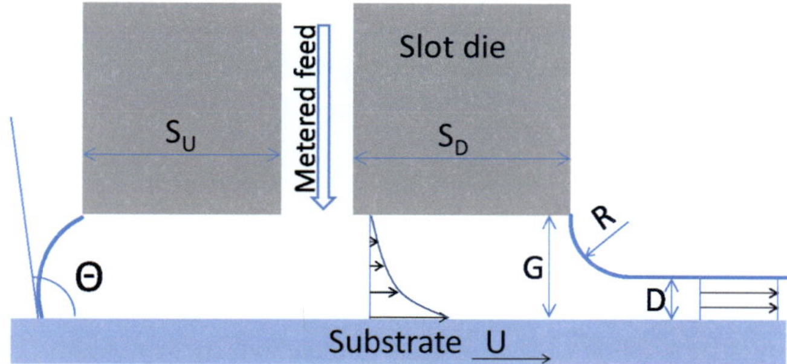

Figure 5.7: Schematic cross section sketch of a slot-die coating process. Velocity profiles for a fully developed flow are indicated by black arrows.

The process is stable when the capillary forces, that are holding the fluid inside the gap, are larger than the viscous forces, acting in the direction of the moving web. This equilibrium was described as a function of the dimensionless gap width by Ruschak (Ruschak 1976) in the Visco Capillary Model (VCM). Carvalho (Voigt, et al. 2012) assumed a minimal radius R of a cylindrical downstream meniscus of (G-D)/2 and neglected the forces on the upstream meniscus, giving:

$$Ca = 0.65 \cdot \left(\frac{2}{G^*-1}\right)^{3/2} \text{ with } G^* = \frac{G}{D}. \tag{5.9}$$

Based on the same model, but including capillary forces at the dynamic contact line of the upstream meniscus, Lee, et al. (Lee, et al. 1992) proposed:

$$Ca = 0.64 \cdot \left(\frac{1+\cos(\theta)}{G^*}\right)^{3/2}, \tag{5.10}$$

as a model for the critical Capillary number. The dynamic contact angle Θ depends on fluid properties as well as coating speed and is here calculated by an empirical approximation from Kistler and Schweizer (Kistler and Schweizer 1997):

$$\theta = 541 \cdot Ca^{0.22} \cdot \sigma^{0.33} \cdot \eta^{-0.04}. \tag{5.11}$$

Material properties were taken from the data presented in Chapter 4. For shear thinning inks, the flow in the gap was assumed to be a fully developed Couette flow, defining the shear rate ($\dot{\gamma}$) as quotient of velocity (u) and gap (G):

$$\frac{du}{dy} = \dot{\gamma} = \frac{u}{G}. \qquad (5.12)$$

Both models are shown in Figure 5.9 and define - for a given speed or Capillary number - the maximum stable dimensionless gap width ($G^* = G/D$); and thus for a given coating gap the minimum achievable wet film thickness. Experimentally it is easier to vary coating speed and ink flow rate than coating gap. Therefore, coating speed and was increased for various volume flow rates until first defects were observed. First defects visualized as ribbing[13] (see Figure 5.8) or constriction of the coating with wavering edges (edge scalloping).

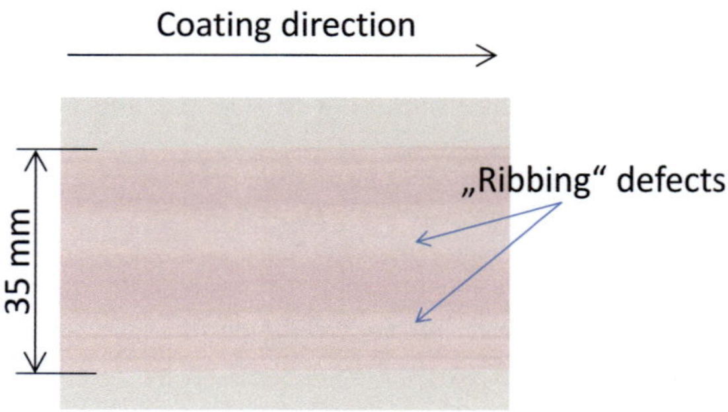

Figure 5.8: Scan of a ribbing type coating defect in a Model-AL (PS in Xyl) film.

The minimal coating speed at which defects occurred first was measured for three material systems, two gap widths and two different die lip designs (the lip width S_D and S_U were reduced from 2 mm to 0.5 mm). Capillary

---

[13] „Ribbing" and „barring" are periodic thickness variation in web- and cross-web-directions, respectively. For a detailed classification of coating defects see: [Gutoff, E. B. and E. D. Cohen (2006). Coating and drying defects: troubleshooting operating problems, Wiley-Interscience.].

Number and dimensionless gap width of these critical operating points are plotted in in Figure 5.9 as symbols. Parameters are summarized in Table 5.1.

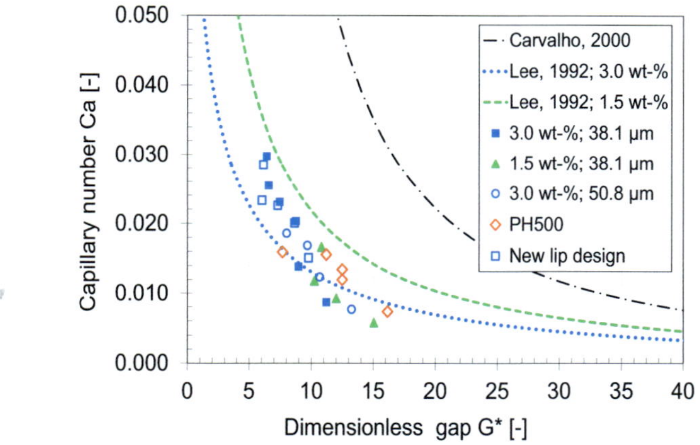

Figure 5.9: Comparison of literature models (dotted and dashed lines) and experimental data (symbols, see Table 5.1) for the stability limit in slot die coating of aqueous and solvent based inks for PSCs. The experimentally observed stability limit is significantly lower compared to the prediction by Carvalho (dot-dashed line, Equation (5.9)) and does not show an impact of viscosity as predicted by Lee (dotted and dashed line, Equation (5.10)).

The visco-capillary model according to Carvalho (Equation (5.3)) is shown as dash-dot line, the model proposed by Lee was computed for a 3.0 wt.-% (dotted line), and 1.5 wt.-% (dashed line) solution of the organic dummy system.

Independent of the dimensional process parameter, homogeneous films are produced at low Ca numbers or low G*, according to experimental data and theory. The Capillary Number predicted by the VCM according to Carvalho (Equation (5.9)) is up to 5 times higher than the experimentally determined values.

Table 5.1:Summary of the parameters used in the experimental data series ind Figure 5.9.

Symbol	Name	Solid content	Gap width	Lip Length
		[wt.-%]	[µm]	[mm]
■	Organic dummy system	3.0	38.1	2.0
▲	Organic dummy system	1.5	38.1	2.0
○	Organic dummy system	3.0	50.8	2.0
◇	PH500	0.6	38.1	2.0
□	Organic dummy system	3.0	38.1	0.5

The model by Lee et al. (Equation (5.10)) predicts Capillary Numbers in the same order of magnitude as the experimental data but a dependence of film stability on ink properties is not observed experimentally. Therefore, a modification of the visco-capillary model was used for an empirical fit of the experimental data:

$$\text{Ca}_{Crit} = m \cdot \left(\frac{2}{G^*-1}\right)^{3/2} \text{ with } m = 0.128,$$

(5.13)

and is shown as solid line in Figure 5.10.

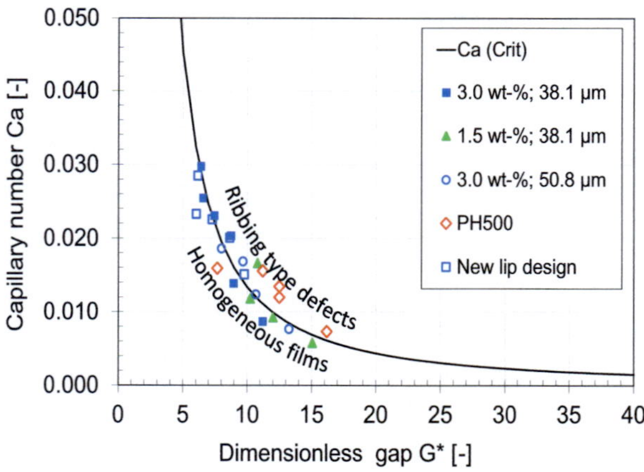

Figure 5.10: Dimensionless coating window for slot-die coating of aqueous and solvent based inks for PSCs. A good fit of the experimental data (sym-

bols, see Table 5.1) was achieved by a modified visco-capillary model (solid line, Equation (5.13)).

Above Ca_{Crit} coating defects occurred, independent of viscosity, surface tension, die design, line speed or coating gap. The diminished stability compared to the VCM can be explained by the low gap widths and viscosities used here. A local deviation in gap width of few micrometers can be caused by inhomogeneous substrate thickness or mechanical precision of the die lip. In conventional coating processes this effect is insignificant since gap widths are in the order of hundreds of micrometers. However, for a gap width of less than 50 µm, the deviation of few micrometers can result in local destabilization of the coating flow. Small perturbations caused by impinging air at the upstream meniscus, or inhomogeneity on the substrate or the dye lip and fluctuations in line speed may have a severe effect due to the low viscosity of the liquid capillary bridge. At $G^*>3$ recirculation will inevitably occur, as predicted by the analytical solution in a Couette-Poiseuille channel flow (see e.g. (Willinger 2013)). Though this must not constitute a stability problem, the probability of coating defects increases with increasing G^* for $G^*>3$ and the coating gap should thus be selected as low as possible. An extension of the coating window by applying a differential pressure over the fluid bridge may increase the maximal stable G^* at a given coating speed but may also result in reduced film quality.

When these considerations are taken into account, slot-die coating can be used to coat the active and hole conducting layer in polymer solar cells with high homogeneity. The stability window in Figure 5.9 was successfully applied and confirmed by producing large area coatings of P3HT:PCBM, P3HT:QD and PEDOT:PSS as well as working solar cells on flexible substrate with a pilot-scale coating line.

In comparison to other coating methods, film thickness and process stability can be determined before an experiment. The process can thus be scaled up and transferred directly to other material systems, which is especially appealing for the rapidly changing material combinations in the field of polymer based solar cells.

5.4. Impact of coating method on film properties

A comparison of cross-profiles for different coating methods is shown in Figure 5.11. As these measurements were made on glass to ensure a flat baseline, a comparison with gravure coated films was not possible. Photoactive films produced by knife, slot-die, and spin coating show a similar macroscopic homogeneity. For knife and slot-die coating, film thickness varied by less than 6 % in cross coating direction, for a coating width of 50 mm and a film thickness of 80 nm.

The same tendency was observed for P3HT:PCBM and PEDOT:PSS films (see Appendix 9.10.6). As discussed above (Subsection 5.3.3), thin homogeneous films could not be produced with the spray coating setup used here.

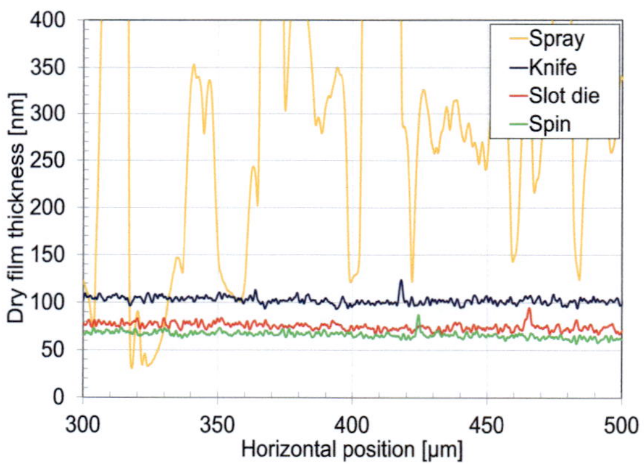

Figure 5.11: Cross-profile measurements of knife, slot-die, spray and spin coated P3HT-M:QD photoactive films (Dektak 3M, Veeco).

A statement about the actual surface topography requires a lateral definition in the same order of magnitude as the film thickness. Due to their high roughness, spray coated films could not be analyzed by AFM. Figure 5.12 shows the surface topography of the hole-conductive and photoactive layers for knife, slot-die, and spin coated samples, only.

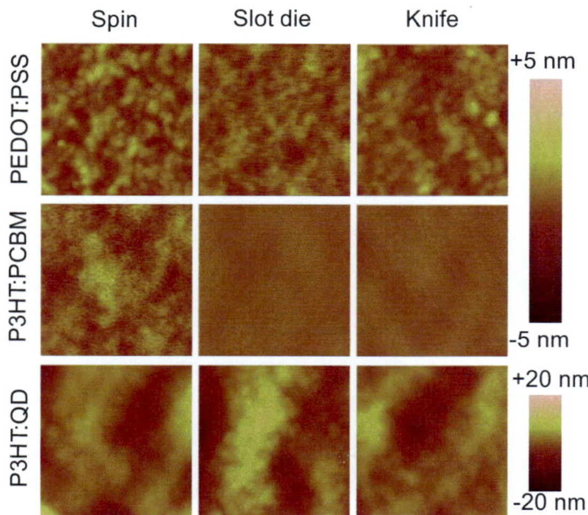

*Figure 5.12: Atomic force microscopic (AFM) pictures of a 500 x 500 nm²
section of hole conducting layers (top row), organic (middle row), and hybrid (bottom row) photoactive layers prepared by spin (left), slot-die (middle), and knife (right) coating.*

Especially for the organic photoactive film (P3HT-P200:PCBM) the surface topography is more pronounced for spin coated films compared to
knife and slot-die coated films. This may be due to constant shearing of the
spin coated film which inhibits leveling of the film during drying or due to
changes of the film morphology. This effect is less pronounced for the
hole-conductive layer and insignificant for hybrid layers where the overall
roughness is much higher ($Rq_{P3HT:QD}$~6-10 nm; $Rq_{P3HT:PCBM}$~0,6 nm) due to
agglomeration of particles within the film. No pinholes or defects were ob-
serves for slot-die, knife and spin coated films.

AFM topography often correlates to the opto-electric properties of a film as
a rougher film may indicate a higher degree of crystallization (see e.g.
(Yang, et al. 2005; Schmidt-Hansberg, et al. 2009b)). Figure 5.13 shows
light absorption of the active layer as a function of wavelength. P3HT ab-
sorbs light above ~400 nm; the shoulders at 550 nm and 600 nm are an indication for its cristallinity. Below 400 nm, the light absorption is dominat-

ed by PCBM (with its peak at 350 nm) or semi-conducting nanoparticles (with increasing absorption towards lower wavelengths) (see Subsection 3.3.2 for spectra of pure substances). For all methods, the crystallization of P3HT is suppressed in hybrid films compared to polymer fullerene films.

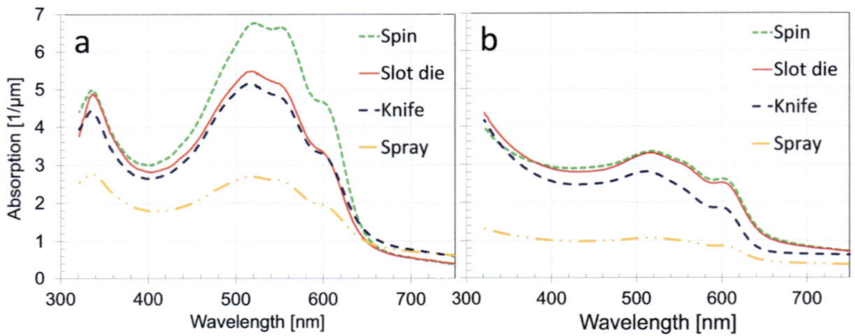

Figure 5.13: Specific absorption spectra of photoactive films for polymer-fullerene (P3HT-M:PCBM; a) and polymer-quantum dots (P3HT-M:QD, b) solar cells produced by spin (dotted), slot-die (solid), knife (broken), and spray (dot-dash) coated films.

Spray coated films show less absorptions at all wavelengths. This is due either to a local variation in film thickness or to low cristallinity caused by rapid drying during atomization and deposition with dry nitrogen. In the lower wavelength regime, the absorption curves differ only slightly for all other coating techniques.

Although spin coated films dry faster than knife and slot-die coated films, they show significantly higher absorption at wavelengths above 500 nm. This indicates that drying kinetics is not the only factor determining the degree of crystallization. Functional hybrid and polymer-fullerene solar cells were produced by all methods (see Section 7.2).

The coating method has a significant influence on the evolution of nano-morphology and electro-optical properties in polymer solar cells (Wengeler, et al. 2012). With respect to the coating fluid, the main difference between the individual coating methods - apart from the drying rate - is the intensity of viscous shear during coating. Whereas the directed shear-rate of droplets within a spray is zero, shear rates of ~100-10 000 1/s are present in knife and slot-die coating (Equation (5.12)), with slot-die coating having the longer residence time in the shear field. Depending on the position on the substrate, the shear rate during spin coating is even higher (~100 000 1/s; Equation (5.4)) and continues throughout the drying period. Thus, the increase in absorption, observed here, could be the result of polymer alignment caused by the coating method's higher shear intensity. To verify this hypothesis, more data is needed for a single coating method under identical drying conditions and varying shear intensity.

5.5. Shear-rate dependence of functional films

As slot-die coating is a pre-metered coating method, process parameters, such as coating speed or coating gap distance, can be varied without changing the wet film thickness. Two different methods to vary the shear-rate during coating were applied here.

1. An increase in coating speed results in an increased shear-rate within the die's slot as well as in the gap between die lip and the moving substrate. The volume flow was adjusted to produce an identical film thickness (Figure 5.14.a).
2. An increase of the slot width results in a decrease of viscous shear within the die; whereas the viscous shear in the coating gap remains constant (Figure 5.14.b).

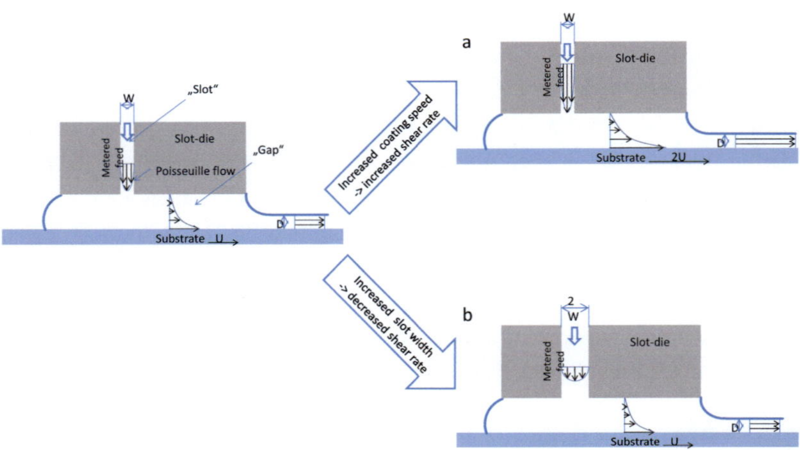

Figure 5.14: Schematic sketch of two methods to vary viscous shear in the slot-die coating process, without changing the wet film thickness. An increase in coating speed (a) increases the shear-rate both within the die as well as in the capillary gap between die lip and substrate. A change in slot width (b) changes only the shear-rate within the die.

Films, coated under variable conditions, were characterized according to their function within the solar cell. The light absorption of photoactive material is presented for two solvent compositions: pure chloroform and a chloroform-chlorobenzene mixture. Subsequently the specific conductivity of transparent polymer electrodes with two solvent compositions and reflective silver nanoparticle electrodes is discussed.

5.5.1. Light absorption of photoactive films

Figure 5.15 shows the specific absorption spectra (normalized with dry film thickness) of P3HT-P200:PCBM photoactive films, slot-die coated from a chloroform solution for six different coating speeds. As illustrated in Figure 5.14.a, the feed flow rate was adjusted to produce films with identical thickness whereas the shear-rate within the die slot and coating gap increases with speed. As a general tendency, films coated at a higher speed exhibit a higher specific light absorption. Surface roughness was similar for all samples with 3.1 nm < Ra < 4.8 nm.

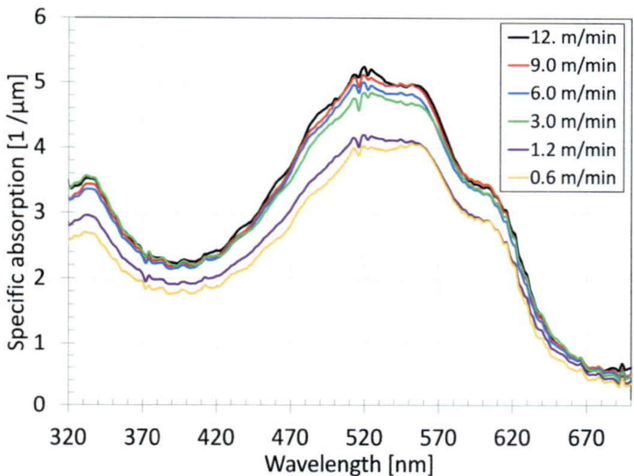

Figure 5.15: Light absorption spectra of P3HT-P200:PCBM (1:0.8 by wt.) photoactive layers, slot-die coated from chloroform at different coating speeds from a 10.8 mg/ml solution. The pump flow rate was adjusted to produce a constant film thickness of 40 nm. Absorption increases with coating speed, indicating a shear-rate induced crystallization.

To compare these results to other coating ink compositions, the specific absorption was averaged between the wavelength of 320 and 720 nm. Figure 5.16.A shows a second ink composition with 50 v.-% CB in addition to the pure CF ink. For both inks, the average absorption increases with a declining slope towards higher coating speed. The effect is more pronounced for the low boiling point solution of pure CF. Figure 5.16.b shows a variation of slot width as illustrated in in Figure 5.14.b at a constant coating speed of 3 m/min. The light absorption shows a decreasing trend from 2.8 μm at 50 μm slot width to 1.7 /μm at a slot width of 200 μm.

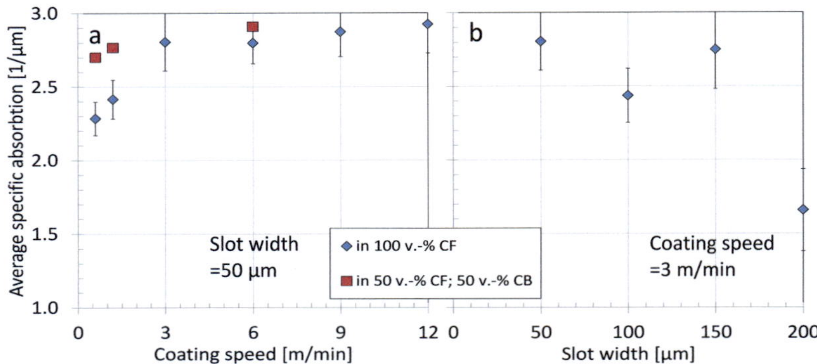

Figure 5.16: Averaged specific absorption of P3HT-P200:PCBM (1:0.8 by wt.) photoactive layers, slot-die coated at different coating speeds (a) and different slot width (b). Absorption decreases towards lower coating speed and higher slot width. The decrease towards low coating speed is more pronounced for a solution containing 100 v.-% Chloroform (diamonds).

The increase in average specific absorption at higher shear rate, due to higher coating speed or smaller slot width, confirms the hypothesis of a shear-induced crystallization of the photoactive material. This result is in good agreement with the results obtained for different coating methods (Section 5.4), where specific light absorption was observed to be higher for coating methods with higher shear intensity. The increased absorption of the low-boiling CB:CF solution is caused by the slower drying rate of this ink, an effect that has been studied in detail before (Schmidt-Hansberg, et al. 2009a).

5.5.2. Conductivity of transparent polymer electrodes

The same coating experiments were conducted for transparent polymer electrode films. Figure 5.17 shows the specific conductivity of PH1000 films, slot-die coated at speeds from 0.6 to 18 m/min. The conductivity of the annealed samples (squares) decreases by 25% from its highest value at 0.6 m/min. The value at 9 m/min (unfilled square) stands out from this general tendency, most likely due to a local coating defect. The blue diamonds represent samples that were characterized prior to annealing. At speeds higher or equal to 12 m/min, the substrate was stopped for approximately 5 min because the drier capacity was insufficient to continuously dry the films. Whereas no significant change is observed for the trend of the annealed samples, the conductivity of the as prepared sample increases discontinuously towards the value of the annealed samples. Due to the longer residence time within the drier, the residual water content is lower for these samples and thus leads to an increased conductivity. The fact that no similar dis-continuous change is observed for the annealed samples indicates that the impact of the residence time within the drier, which changes due to the constant drier length from 3.3 min at 0.6 m/min to 0.2 min at 7.5 m/min, has no significant effect on the annealed samples.

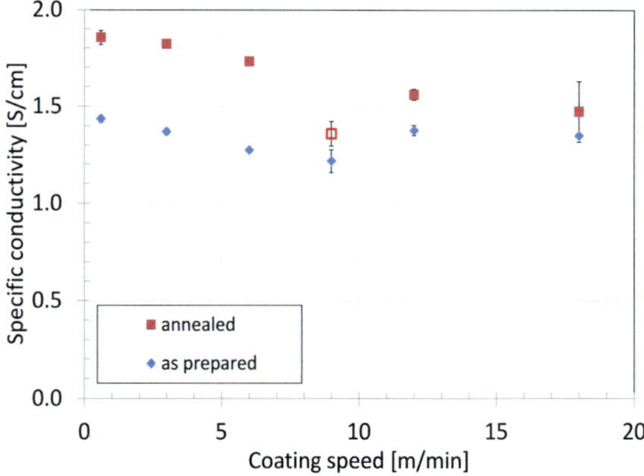

Figure 5.17: Specific conductivity of PH1000 films as a function of slot-die coating speed at a slot width of 50 μm. PH1000 was diluted with MeOH in a volume ratio of 1:0.25 to enhance wetting properties. Conductivity increases after annealing but decreases with coating speed. At speeds >12 m/min, the substrate had to be stopped within the drier to allow for complete drying, resulting in increased values for the as prepared samples due to an additional annealing time inside the dryer.

The magnitude of the specific conductivity of PH1000 films prepared from aqueous dispersion is in good agreement with literature (Xia, et al. 2012). An increase in conductivity by up to three orders of magnitude can be achieved by acid treatment of the dry film (Xia, et al. 2012) or by addition of a high boiling additive to the coating ink.

The same tendency – a decrease in specific conductivity with coating speed – can be observed for films coated from a PH1000:DMSO:TX100 dispersion (Figure 5.18.A). The addition of 5 v.-% DMSO required an additional 0.1 v.-% TX100 to enable sufficient wetting. Figure 5.18.b shows a variation of the die's slot-width as illustrated in Figure 5.14.b. Specific conductivity decreases as a function of slot width, however the decrease lies within the same order of magnitude as the statistic experimental error, indicated by the standard deviation (error bars).

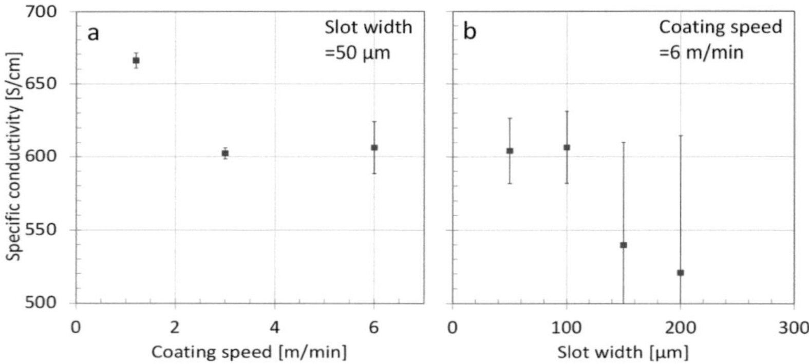

Figure 5.18: Specific conductivity of PH1000 films as a function of slot-die coating speed (right) and slot width (left). Conductivity was enhanced by the addition of 5 v.-% DMSO and wetting properties were adjusted by the addition of 0.1 v.-% Triton X100.

An increase in crystallization with higher coating speed should result in an increased conductivity (Takano, et al. 2012) and was not observed here. In contrast, the specific conductivity decreased with coating speed more pronounced for the ink without DMSO. The small improvement of PEDOT:PSS conductivity with decreasing slot width (Figure 5.18.b) may be caused by an increased shear-rate, but is insignificant compared to the decrease of conductivity at higher coating speed (Figure 5.17 and Figure 5.18.a).

5.5.3. Conductivity of reflective silver nanoparticle electrodes

Figure 5.19 shows the conductivity of silver nanoparticle films as a function of coating speed. Whereas the conductivity exhibits no clear tendency, the standard deviation, plotted as error bars, increases towards higher coating speed indicating a decrease in film homogeneity.

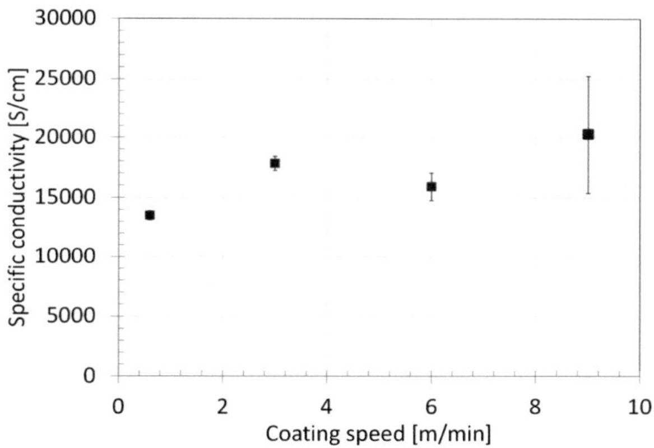

Figure 5.19: Specific conductivity of Ag-nanoparticle films (350 nm thickness) as a function of coating speed. Standard deviation of six measurements is illustrated by error bars and increases with coating speed.

A possible explanation for this phenomenon may be a lower film quality due to the fact that the operating point approaches the stability limit at higher speeds. Figure 5.20 shows the operating points of the three material systems within the dimensionless coating window presented in Subsection 5.3.5.

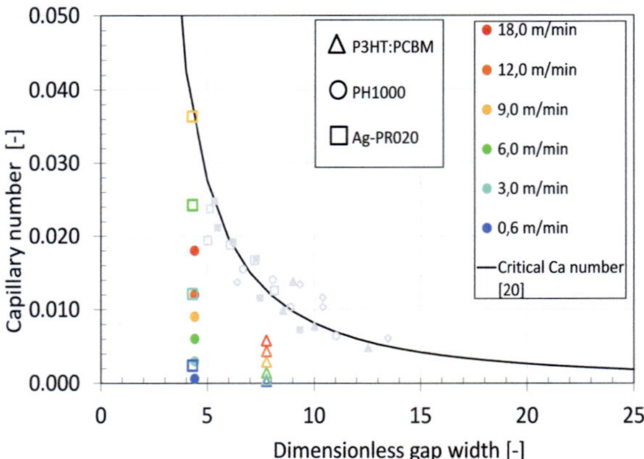

Figure 5.20: Position of the operating points within the stability window for slot-die coating of organic photovoltaic (OPV) material (see [22]) for photoactive material (triangles), high conductive polymer (circles) and silver nanoparticles (squares). Due to the their higher viscosity compared to the photoactive material inks, the operating points for PH1000 and Ag-PR020 coatings approach the stability limit towards higher coating speeds.

Whereas for photoactive films the Capillary number remains at low values (e.g. 0.0044 at 9 m/min), primarily due to the low viscosity, it is close to the stability limit for PEDOT:PSS and Ag-PR020 (e.g. 0.009 and 0.036, respectively; at 9 m/min). The RMS surface roughness of all coatings is plotted as a function of coating speed in Figure 5.21. As expected from the position of the operating point within the stability window, the surface roughness of electrode films increases at high coating speed.

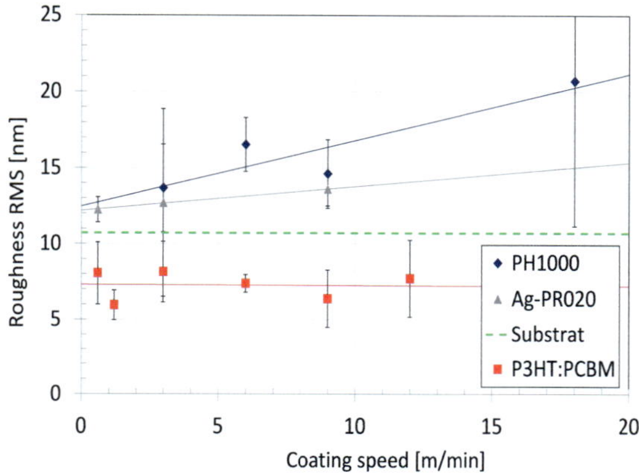

Figure 5.21: Surface roughness determined by profiler for PH1000, AgPR020 and P3HT-P200:PCBM films as a function of coating speed. The PET substrates roughness is given as reference.

By employing a vacuum chamber, higher coating speeds may be possible (Jakubka, et al. 2013) but it remains to be seen whether sufficient film quality can be achieved beyond the limit presented here.

The different results for the three material systems can be attributed to their different nature. A shear-induced crystallization, caused by an alignment of polymers in the shear field of the coating flow, can only be expected for a polymer solution. Correspondingly, a significant increase in crystallization related properties with increased shear rate was observed for the polymer-fullerene solution, only. The dispersions: PH1000 and PR020 were also affected by the choice of process parameters though whether a change of morphology is caused by e.g. a break-up of agglomerates in the shear field cannot be proven. More important for both electrode systems was the position of the operating point within the dimensionless coating window, resulting either in lower conductivity (PH1000) or higher deviation of the conductivity values (Ag-PR020).

6. Technical drying of functional films

6.1. Introduction and state of the art

The influence of drying kinetic on the morphology of spin coated photoactive films was observed by Li, et al. in 2005 (Li, et al. 2005). Mihailetchi showed that the effect of drying rate is conserved after annealing by showing that slowly dried samples exhibit higher efficiency compared to fast dried samples with subsequent annealing (Mihailetchi, et al. 2006). Schmidt-Hansberg used a laboratory scale roll-to-roll compatible knife coating setup and in-situ grazing incident x-ray diffraction (GIXRD) to monitor polymer and fullerene crystallization during film drying (see e.g. (Schmidt-Hansberg, et al. 2009a; Schmidt-Hansberg, et al. 2009b; Schmidt-Hansberg, et al. 2011b; Schmidt-Hansberg 2012)). The experiments were conducted under isothermal conditions at heat transfer coefficients of less than 10 W/(m²·K) (Schmidt-Hansberg 2012). He concluded that morphology of P3HT:PCBM blends is improved by low isothermal drying temperature, and slow drying rates, supported by high boiling point solvents. The rate of heat and mass transfer, varied by changing air velocity in the drying channel, had only a moderate influence. The same effect can be observed for hybrid photoactive P3HT:QD blends as documented in Appendix 9.10.7.

So far, experiments were conducted under laboratory conditions. In Section 6.3 the impact of transfer coefficients, air temperature and solvent composition under technical conditions is discussed and compared to laboratory results.

Though a slow drying rate and low temperature may be beneficial to film morphology, they also provide time for de-wetting processes or the amplification of film instabilities. According to Sharma and Ruckenstein (Sharma and Ruckenstein 1990), all fluids that exhibit a contact angle $>0°$ are metastable below a critical film thickness. In this regime, a critical defect radius, which scales with film thickness, defines whether a defect propagates. Thus, films that were coated under meta-stable conditions may become instable during drying since surface tension and surface energy change with time and film composition. The impact of solvent boiling point and drying rate on the stability of films in roll to roll PSC production has been reported by various authors (Kopola, et al. 2010; Schrödner, et al. 2012; Voigt, et al. 2012), but has not been investigated further.

To optimize the drying process on a technical scale, the improved morphology has to be balanced against a reduction in film stability for slow drying rates. First experiments are published in Section 6.4.

6.2. Specific experimental details

The films in **Section 6.3** were produced by knife coating on the Batch Coater or Pilot Coating Line depending on solvent composition.

For inks with pure solvents the conditions in a technical process were emulated on the Batch Coater (Subsection 6.3.1). The substrate temperature was set to the calculated thermodynamic film temperature (see Table 6.1). The films were moved back and forth within the ASN-drier at 20 mm/s over a distance of 100 mm for 10 cycles, resulting in a residence time of 100 s inside the ASN-drier.

Experiments with solvent mixtures (Subsection 6.3.2) were conducted on the Pilot Coating Line. A glass substrate was positioned on the continuous PET substrate and dragged by the moving PET substrate. Films were knife coated in Coating-Module-3 (see Subsection 2.3.1) and entered the PSN-drier immediately after coating. The drier temperature was set to 120°C and blower power to 100 % in both segments.

All films in Section 6.3 were produced by knife coating with a gap of 100 μm, a liquid reservoir of 100 μl, and a coating speed of 20 mm/s. The solutions had a P3HT-P200 concentration of 6 mg/ml and a PCBM concentration of 4.8 mg/ml. Light absorption was measured before and after annealing for 10 min at 120°C under ambient conditions.

The P3HT-P200:PCBM films presented in **Section 6.4** were slot-die coated at 3 m/min with a gap width of 38 µm, and a slot width of 50 µm. The PSN-drier was set to 100 % power (in and out) and both segments were set to 120°C.

PH1000 films were knife coated with 100 µl ink, and a coating speed of 5 mm/s and dried under ambient conditions.

The drying simulation in Section 6.4 was simplified, assuming ideal solvent behavior, an adiabatic film with negligible heat capacity, and no diffusional mass transport limitation within the film. The transient component mass balance of a film element was solved numerically by a differential step method:

$$m_i(t + \Delta t) = m_i(t) + \frac{dm_i}{dt}(t) \cdot \Delta t, \qquad (6.1)$$

with

$$\frac{dm_i}{dt} = \beta \cdot \tilde{\rho}(\tilde{y}_F - \tilde{y}_\infty), \qquad (6.2)$$

where m_i is the mass of component i per unit area.

The convective and radiative heat transfer coefficients were set to 50 W/(m²·K) and 5 W/(m²·K), respectively. Material properties used in these calculations are tabulated in Appendix 9.7.

6.3. Technical drying of photoactive layers

In contrast to the isothermal conditions in the laboratory set-up used by Schmidt-Hansberg, film temperature and solvent properties are closely related in a technical process. The film assumes its "Beharrungstemperature", which is defined by the steady state equilibrium of heat transfer into the film and evaporative cooling during the first drying period. For a simple approximation, substrate and film may have a negligible heat capacity, temperature and concentration are constant throughout the film, and the solvent mol-fraction in the drying air is small. Radiation as well as the Stefan and Ackerman correction (see e.g. (Stephan 1994)) are neglected. The Beharrungs-temperature differs from the well-known Wet-Bulb-

Temperature (see e.g. (Walker, et al. 1923)) due to the different areas associated with mass and heat transfer. Whereas heat is transferred into the film from bellow and above (film and substrate side), mass transfer occurs at the film's surface, only (resulting in the factor 2 in Equation (6.3)). Under these assumptions, the energy and mass balance of a differential film element yield:

$$\beta \cdot \tilde{\rho} \cdot (\tilde{y}_F - \tilde{y}_\infty) \cdot \Delta \tilde{h}_v = 2 \cdot \alpha (T_\infty - T_F), \tag{6.3}$$

where β and α are mass and heat transfer coefficient, $\Delta \tilde{h}_v$ the solvent's molar enthalpy of evaporation. Solvent mol-fraction and temperature are: $\tilde{y}$ and T with the subscript indicating the position either at the film surface (lower index F) or in the bulk drying air (lower index ∞). $\tilde{y}_F$ was computed as vapor pressure at film temperature divided by ambient pressure and $\tilde{y}_\infty$ was assumed to be zero for all organic solvents. Heat and mass transfer coefficients are correlated according to Lewis law:

$$Le^{1-n} = \left(\frac{Sc}{Pr}\right)^{1-n} = \left(\frac{\lambda_{Heat}}{\tilde{\rho}\tilde{c}_p \delta_{i,Air}}\right)^{1-n} = \frac{\alpha}{\beta} \cdot \frac{1}{\tilde{\rho}\tilde{c}_p}, \tag{6.4}$$

where n is 1/3 for laminar and 0,42 for turbulent flow and Le, Sc and Pr are dimensionless groups called Lewis, Schmitt, and Prandtl Number, respectively. The dimensionless groups can be expressed as a function of the properties of the drying air: heat conductivity (λ_{Heat}), molar density ($\tilde{\rho}$), heat capacity ($\tilde{c}_p$), and the diffusion coefficient ($\delta_{i,Air}$) of solvent "i" in air.

Vapor pressure was calculated from Antoine correlations and diffusion coefficients were computed according to Fuller. The properties of air were taken from fits of tabulated data for an average boundary layer ture $T_{Boundary} = (T_F + T_\infty)/2$. The relevant material properties are documented in Appendix 9.7. Thus Equation (6.3) and (6.4) define the film temperature as an implicit function of air temperature and solvent properties:

$$(\tilde{y}_F - \tilde{y}_\infty) \cdot \Delta \tilde{h}^v = 2 \cdot \tilde{c}_p Le^{1-n}(T_\infty - T_F). \tag{6.5}$$

Film temperatures for typical solvents calculated from Equation (6.5) are presented for three different drying air temperatures in Table 6.1.

Table 6.1: Example for film temperature calculated from Euation (6.5) for three drying air temperatures and typical chlorinated and non-chlorinated solvents used for polymer solar cells.

Air temperature:	80 °C	100 °C	120 °C	Boiling point
Solvent		Film temperature [°C]		[°C]
o-Dichlorobenzene (DCB)	70	83	94	179
Indane (IND)	70	83	93	178
m-Xylene (XYL)	58	67	76	144
Chlorobenzene (CB)	55	64	72	132
Toluene (TOL)	46	54	60	111
Chloroform (CF)	20	25	30	61

For films dried at the same air temperature, film temperature might differ up to 64°C depending on the solvent. The difference between air and film temperature of high boiling point solvents (DCB, IND) is less than 30°C. It increases for intermediate boiling point solvents (CB, XYL) and can reach up to 90°C for low boiling points solvent (CF, TOL). Toluene exhibits the lowest boiling point of all non-halogenated solvents with sufficient solubility for P3HT:PCBM (Machui, et al. 2012) and is thus grouped as low boiling point solvent together with the much more volatile Chloroform.

The assumption that heat capacity of the substrate and film can be neglected is not valid for typical pilot scale experiments. A 96 µm PET foil, introduced from ambient temperature (20°C) into a drier with an air temperature of 120°C and a heat transfer coefficient of 30 W/(m²·K) reaches 80% of the final temperature increase (80°C) after 38 seconds. Thus especially for the low boiling point solvents the estimated drying times are in the same order of magnitude as the transient heating time and the exact film temperature results from an energy balance of both. In the next section the film temperatures were set to the values from Table 6.1 in order to investigate the influence of film temperature independent of the substrate type and thickness.

6.3.1. Drying temperature and heat transfer coefficient

Light absorption spectra for P3HT-P200:PCBM films dried from toluene solution at different drying air temperatures are presented in Figure 6.1.

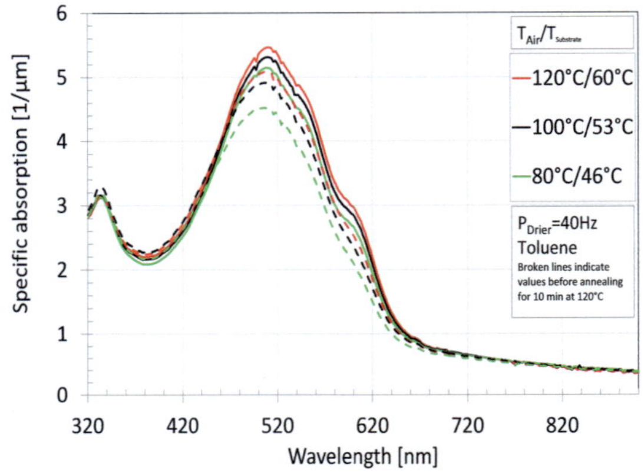

Figure 6.1: Variation of drying air temperature for films dried at wet bulb temperature from a Toluene solution. The light absorption of the P3HT-P200:PCBM films increases in the polymer dominated regime with air temperature.

Whereas the fullerene-peak does not change significantly, the absorption at wavelengths characteristic for crystalline polymer increases with drying air temperature.

The increase is more pronounced for as-prepared samples (broken lines) compared to annealed samples (solid lines), indicating that this may be caused by thermal annealing of the dried films. The absolute difference in absorption lies within the statistical deviation of the measurement (~10%).

In contrast, Schmidt-Hansberg observed an increase in polymer crystallinity toward lower temperatures (Schmidt-Hansberg 2012). The impact of a lower drying rate due to lower film temperatures may be relevant at low transfer coefficients and temperatures, only. The light absorption at the lower temperature and variable blower frequency is shown in Figure 6.2. The heat transfer coefficient is proportional to the frequency of the blower (see Appendix 9.5.2). Exact values are currently being measured and will be published by Jaiser. The highest absorption was measured for the highest frequency and thus highest drying rate. The absolute difference of absorption before and after annealing is small and no clear tendency can be observed.

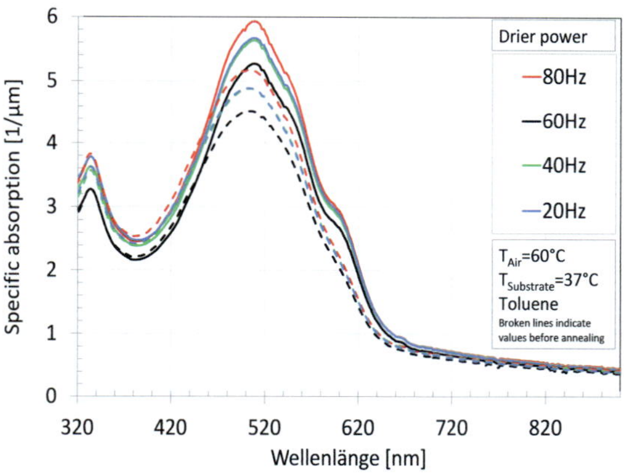

Figure 6.2: Light absorption spectra for P3HT-P200:PCBM films dried from Toluene at an air temperature of 60°C and variable blower frequency.

To compare the impact of drier settings for different solvents the average absorption between 320 and 720 nm was computed. The same knife coating parameters were used in all experiments, eliminating a possible impact of the coating process but resulting in different film thicknesses for different solvents. Therefore a comparison of the absolute values for different solvents is not possible and absorption is shown normalized by the average of each solvent series in Figure 6.3.

Though some variation can be observed, especially at high drier frequencies, no overall trend of light absorption as a function of blower frequency and thus drying rate is apparent. In concurrence, Schmidt-Hansberg reported only a moderate impact of heat transfer coefficients (Schmidt-Hansberg 2012).

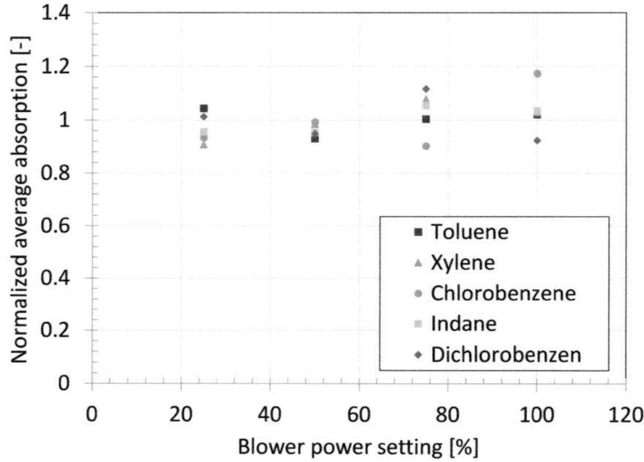

Figure 6.3: Normalized average absorption at different blower settings for typical solvent systems used in OPV. The air temperature was set to 60°C for IND, DCB, und TOL and 100°C for CB und XYL. Substrate temperature was set to the corresponding wet bulb temperature. No general trend of absorption as a function of blower frequency can be observed.

In summary, the impact of drying temperature and heat transfer coefficient on light absorption is limited under technical conditions. To reach the region where lower film temperatures can improve the morphology of the bulk-heterojunction, as reported by Schmidt-Hansberg, the drying rate has to be reduced to values below those possible with the technical drying equipment used here. A possible solution for a continuous process was proposed by Park (Park, et al. 2010), who covered the film with a silicon sheet to reduce the drying rate.

6.3.2. Solvent composition

In comparison to drying temperature and heat transfer coefficients, a change of solvent affects both, drying rate and solubility. In general, cell efficiency can be improved by using high boiling point solvents instead of low boiling point solvents and chlorinated instead of non-chlorinated solvents (see e.g. (Schmidt-Hansberg, et al. 2012)). Since high boiling point solvents require higher drying temperatures and longer drying times, their application increases the capital and operating costs of a process. Schmidt-

Hansberg, et al. proposed the use of a solvent mixture to decrease drying time while sustaining slow drying rates during the critical drying phase at the end of the constant rate period (Schmidt-Hansberg 2012). Figure 6.4 shows specific light absorption spectra of P3HT-P200:PCBM films, knife coated and dried on the pilot coating line, for different Toluene to Indane volume ratios. A significant decrease in light absorption is observed for 90 v.-% and 100 v.-% Toluene, whereas average light absorption at 80 v.-% Toluene is only 3% lower compared to pure Indane.

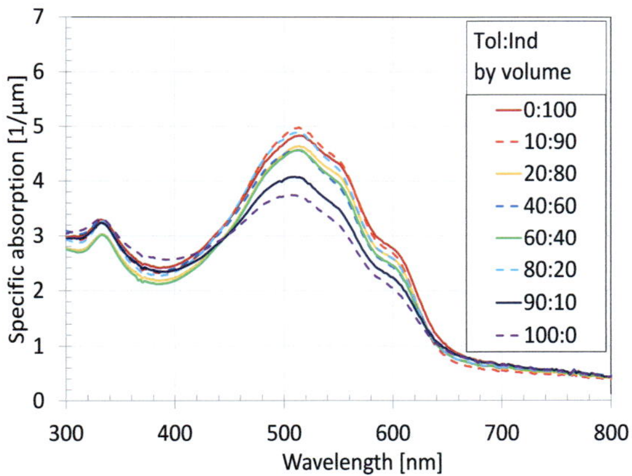

Figure 6.4: Specific light absorption of P3HT-P200:PCBM films knife-coated from solutions with different Toluene to Indane volume ratio. Films were prepared in the pilot coating line and annealed at 110°C for 10 minutes. Light absorption at 100 v.-% and 90 v.-% is significantly reduced compared to the absorption at lower toluene contents.

In contrast to temperature and heat transfer coefficient, solvent composition has a significant effect on light absorption in technically dried photoactive films. The increased process costs, associated with high boiling point solvents, can be reduced by using a mixture of low and high boiling point solvents with only a small reduction in light absorption.

6.3.3. Timescale of solvent sorption in sub-micrometer films

Film temperature and solvent composition have a stronger impact on light absorption than heat and mass transfer coefficients. This fact indicates that morphology is determined in the last drying regime when mass transfer is limited by transfer resistances in the film rather than resistances in the gas phase. At the start of this thesis, drying rate was assumed to be limited mostly by transfer resistance in the gas phase, due to the low film thicknesses of typically less than 200 nm. However, the measurement of a PH1000 film inside a chamber that was flushed with dry nitrogen showed a continuous increase in conductivity for more than 1 h (Figure 6.5).

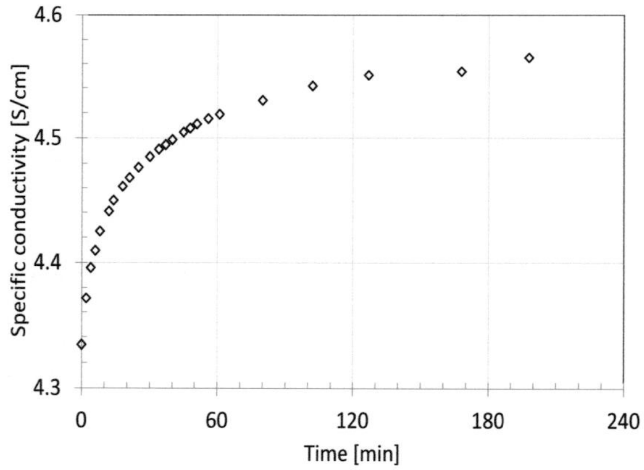

Figure 6.5: Specific conductivity as a function of time of a PH1000 film in a container that is continuously flushed with dry nitrogen.

At equilibrium conditions, conductivity is a function of relative humidity (see Appendix 9.8.1), indicating that this increase is determined by desorption of water at an unexpected long time-scale. Similar characteristic times were observed for thermal annealing of silver nanoparticle films and solvent annealing of PH1000 or photoactive films (see Appendix 0).

The sorption kinetics of solvents in films of sub-micrometer thickness appears to be slower than expected. This effect was observed with quartz crystal microbalance (QCM) measurements for other material systems by

Vogt (Vogt, et al. 2005) and Eastman (Eastman, et al. 2012). For functional films, the sorption kinetic is essential for the design of the process as well as degradation caused by water sorption during operation. A measurement system for investigation of mass transport in nanometer films by QCM is currently being set up by Buss (Buss, et al. 2013).

6.4. Observations on stability of drying films

Beside the impact on morphology, solvent composition and drying rate may also determine film stability and coating quality. The low viscosities and low solid contents, characteristic for functional films, provide little resistance and ample time for destabilizing flows to develop. In polymer solar cells, silver electrodes and electron conducting adaptation layers are typically coated from low surface tension solvents, and are thus not problematic with respect to stability. In contrast, PEDOT:PSS and P3HT:PCBM films are often instable. A detailed investigation of the stability of such films is beyond the scope of this work but some observations made during this work are published below.

6.4.1. Impact of additives and humidity on aqueous dispersions

Coating of aqueous PEDOT:PSS dispersions may be challenging due to the high surface tension of water. Though initial wetting is mandatory to form functional layers for OPV, many wetting related defects occur during the drying process. As the film dries, composition and temperature change and a film that was initially homogeneous may become unstable.

Figure 6.6 shows the impact of humidity and high boiling point additives on the drying kinetic of PEDOT:PSS dispersions. The simplified simulation (see Section 6.2) computes film thickness (blue line), and solvent composition as a function of time. The property models presented in Chapter 4 are used to compute surface tension (red line) and zero shear viscosity (green line). Figure 6.6.a shows a simulation of a PEDOT:PSS dispersion (VPAI40.83 + 15 wt.-% IPA) at low (40 % rH, 22°C) relative humidity. A significant amount of IPA remains within the film until viscosity and SFT increase in the last seconds of the drying process. At a high relative humidity (85 % rH, 22°C; Figure 6.6.B) the mass transport of water is slower and the surface tension approaches the value of pure water, whereas viscosity remains below 10 mPas for more than 30 s. The addition of 5 v.-% DMSO to a dispersion of high conductive PH1000 leads to a still longer drying time, even at low relative humidity (40 % rH, 22°C; Figure 6.6.C).

Though the simplifications of the simulation do not allow for a quantitative evaluation, critical wetting conditions can be determined. A high surface energy and low viscosity for a time span in the order of minutes indicate an undesirable operating point.

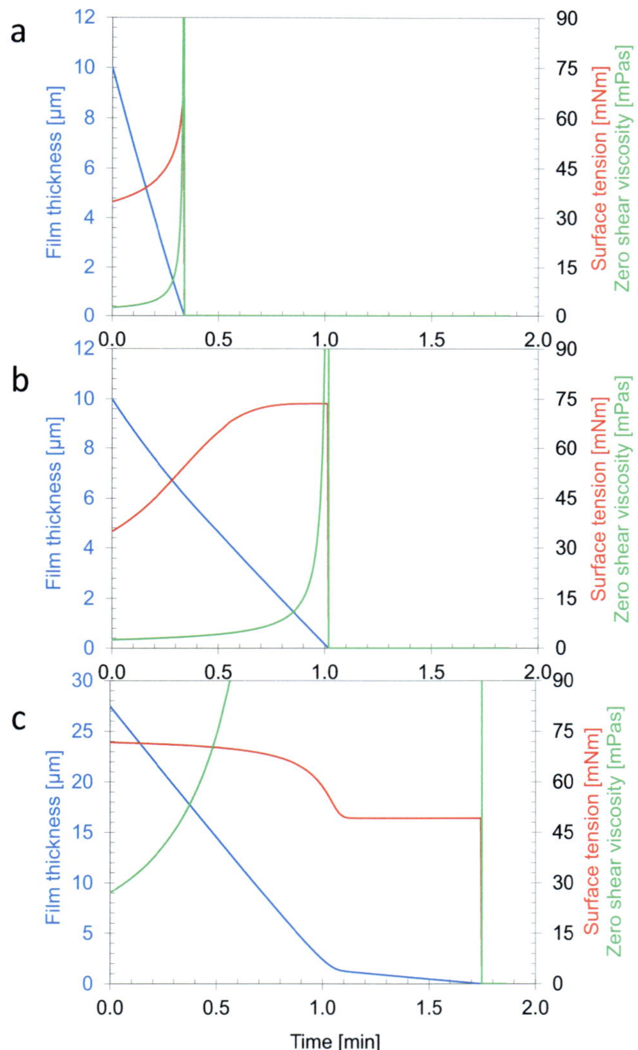

Figure 6.6: Simplified drying simulation of PEDOT:PSS dispersion at an ambient temperature of 22°C. In a and b properties are calculated for a 10µm VPAI40.83 film with 15 wt.-% Isopropanol, at low (40 % rH; a) and high (85 % rH; b) relative humidity. c shows data for a 24 µm PH1000

117

with an addition of 5 v.% DMSO at low humidity (40 % rH). Drying is slowed down by higher humidity and the addition of high boiling point DMSO, resulting in long periods of low viscosity and high surface tension.

Whereas the surface tension of hole-conducting VPAI40.83 dispersions can be adjusted by adding alcohols, the addition of high boiling point DMSO to improve conductivity of PH1000 films highlights the impact of solvent composition on film stability. Even at low humidity the addition of a surfactant is necessary to ensure wetting.

Figure 6.7 shows a picture of PH1000:DMSO films on a plasma treated glass substrate without surfactant (a) and with surfactant (b). Even though both inks originally formed homogeneous films, the film without surfactant receded into a single droplet during drying.

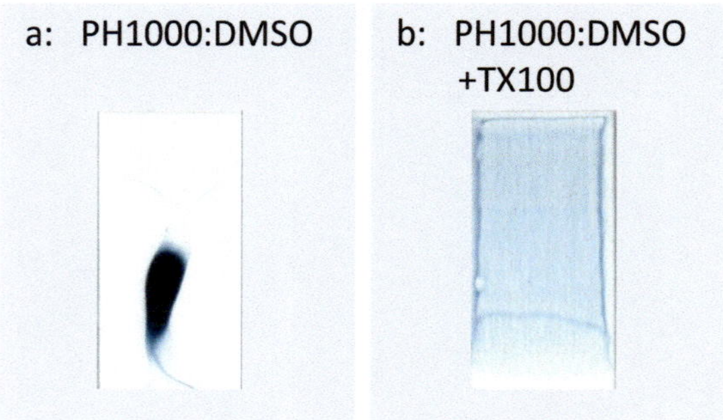

Figure 6.7: De-wetting of a high conductive polymer film on a multilayer OPV stack.

Possible explanations are an increase in surface energy of the plasma treated substrate during drying, or a surface tension driven flow caused by inhomogeneous drying and subsequent local differences in surface tension. Though the problem was solved for this application by the addition of 0.5 wt.-% Triton X100, the interaction of drying and de-wetting kinetic is not fully understood.

6.4.2. Stability of low viscous photoactive films

Similar observations can be made when using high boiling point solvents in P3HT:PCBM inks. Figure 6.8 shows scans of P3HT-P200:PCBM films coated under identical conditions from inks with different solvent. Film homogeneity decreases with increasing solvent boiling point.

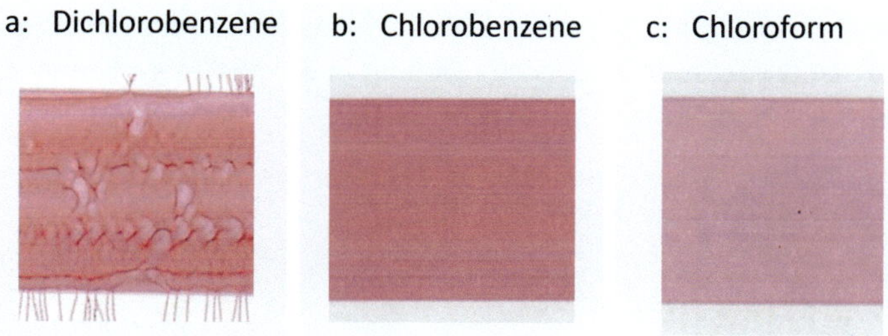

Figure 6.8: Photoactive layer composed of P3HT-P200 and PCBM coated from Dichlorobenzene (a) Chlorobenzene (b) and Chloroform (c), dried under identical conditions in the pilot coating line (120°C, 100% blower setting). Film homogeneity increases with decreasing solvent boiling point.

In contrast to water based PEDOT:PSS dispersions, the surface tension of these inks is less than 40 mN/m and within the wetting envelope of most substrates. Thus, rather than de-wetting, slow drying rates and low viscosity cause inhomogeneity within the film. A possible reason for instability is a gradient in surface tension due to local differences in concentration or temperature. The surface tension of PCBM in Dichlorobenzene is provided in Appendix 9.10.3. Though the absolute change of surface tension is less than 1 mN/m, for concentrations within the solubility limit of PCBM, solutions with high PCBM to P3HT ratio are often unstable. Even small gradient in surface tension may affect film stability due to the low viscosities of solutions with high PCBM and low polymer content (see Subsection 4.2.2).

If it is not possible to change the solvent composition, there are other parameters that can be varied to achieve stable, homogeneous coatings. Figure 6.9 shows three P3HT:PCBM films coated from identical ink. At a wet film thickness of 11.6 μm and a polymer concentration of 6 mg/ml, the coating is obviously instable (Figure 6.9.a). A reduction of wet film thickness significantly increases the film homogeneity due to the reduction of drying time. At constant solid content a reduction in wet film thickness is proportional to dry film thickness (Figure 6.9.b). An identical dry film thickness (as in Figure 6.9.a) can be achieved, by increasing solid content and decreasing wet film thickness by an identical factor. At a polymer concentration of 12 mg/ml, and a wet film thickness of 5.8 µm, viscosity is increased and drying time is reduced resulting in an even more homogeneous film (Figure 6.9.c).

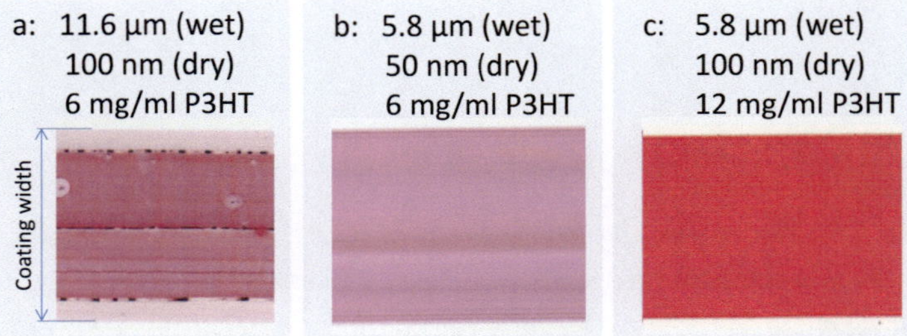

Figure 6.9: Scans of two P3HT-P200:PCBM films slot-die coated and dried under identical conditions (CB:CF – 1:1) but different wet film thickness (a, b) and different solid content (c). The polymer to fullerene ratio was kept constant (1.25).

The length scale at which surface tension can cause lateral flow was defined as Capillary Length (L) by Keddie (Keddie and Routh 2010) as:

$$L = D \left(\frac{\sigma_L}{3\eta\dot{E}} \right)^{\frac{1}{4}},$$

(6.6)

where $\dot{E}$ is the descending rate of the evaporating films surface. A large value for the capillary length will thus result in unwanted convection within the film. Though stability cannot be predicted, Equation (6.6) points out the qualitative influence of process parameters. Film stability decreases with film thickness and surface tension, and increases with viscosity and drying rate. For troubleshooting for a specific film application, stability can be increased by:

- Reducing wet film thickness (D) and increasing viscosity (η), by increasing solid content,
- decreasing surface tension (σ_L) by adding alcohols or surfactants to water based dispersions,
- and increasing rate of evaporation ($\dot{E}$) by using low boiling point solvents or higher film/drying temperature.

The impact of drying kinetic on film stability and the occurrence of drying related defects is at least as important as the optimization of the nano-morphology. For a successful scale-up of a drying process for polymer solar cells a tradeoff between improved stability at fast drying rates and improved morphology at slow drying rates has to be made.

7. Solar cell devices

The focus of this thesis was the implementation of a process for coating and drying of functional films for PSCs and the investigation of both process steps. The logical extension of this work would be the transfer of the insights concerning process stability and film properties to the manufacture of complete cells. For the preparation of cells, up to five individual layers have to be coated on top of each other. Subsequently, the layer stack needs to be prepared for characterization by: cutting or breaking of substrates into sizes that fit into the solar simulator, annealing to remove residual solvents and humidity, evaporation of electrodes (for standard architecture), and transportation to the evaporator and solar simulator. Each additional step introduces a degree of experimental uncertainty so that sample numbers have to be large in order to produce statistically meaningful results. Within the timeframe of this work, a characterization of complete experimental sets with varied process parameters was not possible. Instead, cells were produced exemplarily with hybrid photoactive layers (Section 7.1), by different coating methods (Section 7.2), and with polymer-fullerene photoactive layer in standard and inverted architecture (Section 7.3).

7.1. Hybrid solar cells

Recently, hybrid solar cells with a PCE above 7% have been published (Thon, et al. 2012). These results were achieved by replacing the surface ligands with shorter "linkers". For the material system used here, particle stabilization was done using pyridine as a ligand. The ligands stabilize the particles in the dispersion but remaining ligands can lead to increased electron transfer resistances in the dry film. The highest published efficiency for this system, a PCE of 2.0%, was reported by Zhou, et al. (Zhou, et al. 2010).

A blend of 6 mg/ml P3HT-M and 54 mg/ml quantum dots (QD) was knife coated on a 30 nm PEDOT:PSS film on an ITO-glass substrate. A typical distribution of current density voltage curves for seven cells on one substrate is shown in Figure 7. Cell 1 shows a significantly reduced open circuit voltage and a low parallel resistance, most probably due to short currents through the cell. All other current density voltage curves show a high parallel resistance and no s-shape behavior.

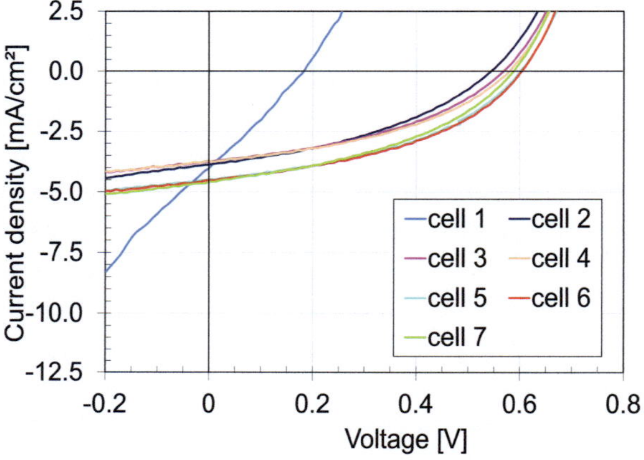

Figure 7.1: Typical distribution of jV-curves for seven hybrid solar cells (P3HT-M:QD) on one substrate. Photoactive and hole conducting (PE-DOT:PSS) layer were knife coated on ITO-glass. Electrode evaporation (7.85 mm²) and characterization was done by Bayer Technology Services.

The characteristic data of the record cell from Figure 7.1 is compared to a spin coated literature value in Table 7.1. Short current density, fill factor and efficiency are higher for the literature record cell, which may be due to an optimized post processing and a much larger sample volume.

Table 7.1: Comparison of hybrid solar cells prepared by different coating methods.

Coating method	Electrode	V_{oc} [V]	j_{sc} [mA/cm²]	FF [-]	PCE [%]
Spin (Zhou, et al. 2010)	Al	0,62	-5,8	0,56	2,0
Knife (on ITO-glass)	Ca/Al	0,61	-4,5	0,43	1,18
Slot-die (on ITO-PET)	Al	0,67	-0,28	0,3	0,06

A hybrid solar cell produced by slot-die coating on PET substrate with the pilot coating line is shown in Figure 7.2. The lower efficiency (see Table 7.1) was probably due to problems with the patterning of the ITO-PET substrate for the Bayer layout (see Section 3.2) and due to contamination and mechanical stress caused by contact of conveyor rolls to the top of the coating.

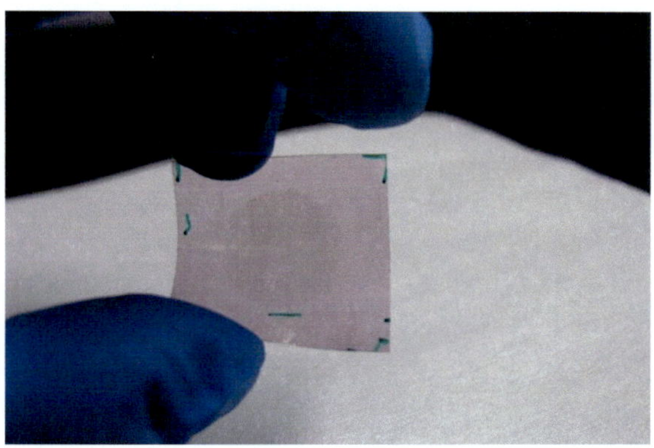

Figure 7.2: Photograph of flexible ITO-PET substrate slot-die coated with a 30 nm hole conduction layer (VPAI4083) and 50 nm hybrid photoactive layer (M-P3HT:QD).

The coating experiments showed good film quality for the P3HT-M:QD films with a standard deviation of 2.8 nm for films with a dry film thickness of 40 nm. The film homogeneity was promoted by higher viscosity of the P3HT-M:QD ink compared to P3HT:PCBM ink, which led to a higher film stability during drying.

Almost identical open circuit voltages and similar order of magnitude in fill factor indicate that a scale-up from laboratory scale spin coating to a continuous coating method is feasible. This result, published in 2011 (Wengeler, et al. 2011) is the first report of a bulk-hetero-junction hybrid solar cell prepared with a large scale coating method.

7.2. Cell performance for different coating methods

In Section 5.4, the coating methods slot-die, knife, spray, and spin coating were compared using the same batch of coating ink that was used to prepare the solar cells in this section. Though cell performance shows strong statistical variation, working solar cells with polymer-fullerene and polymer-quantum-dot photoactive layers were produced by slot-die (SDC), knife (KNC), spray (SRC), and spin (SNC) coating. Gravure coating was not investigated here because the gravure coating system was still under development at the time of the experiments.

Figure 7.3 shows current density voltage curves for polymer-fullerene (Figure 7.3.a) and hybrid (Figure 7.3.b) photoactive layers. In Figure 7.3 only the cells with highest power conversion efficiency for each coating method are plotted. A typical value for the open circuit density of P3HT-M:PCBM cells is ~0.6 V (see e.g. (Brabec 2008)). The low open circuit voltage for all methods except spray coating indicates a systematic experimental error. A possible explanation may be that water is absorbed by the PCBM phase of the photoactive layer during ambient processing. The atomization with dry nitrogen during spray coating may cause higher V_{OC} for the spray coated samples. In contrast, the hybrid cells showed a V_{OC} that is consistent with literature data for the same material system (Zhou, et al. 2010).

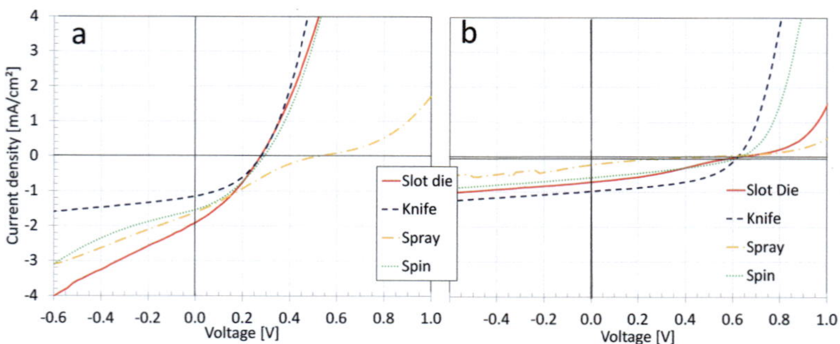

Figure 7.3: Current density voltage curves of polymer-fullerene and hybrid solar cells prepared by different coating methods.

The characteristic data of the jV-curves in Figure 7.3 is documented in Table 7.2. The small slope at V=0 (a high parallel resistance) indicates that the layer is pinhole free and results in a reasonable fill factor of more than 30% for all methods except spray coating. The jV-curves of spray coated films exhibit a strong s-shape behavior which results in a lower fill factor.

Table 7.2: Summary of characteristic data for polymer fullerene and hybrid cells produced with different coating methods and the same material batch used in Section 5.4.

P3HT:PCBM	KNC	SDC	SNC	SRC
η [%]	0,13	0,22	0,16	0,19
V_{oc} [V]	0,28	0,28	0,29	0,53
j_{sc} [mA/cm^2]	-1,15	-1,92	-1,54	-1,62
FF [%]	39,6	32,2	34,8	21,8

P3HT:QD	KNC	SDC	SNC	SRC
η [%]	0,29	0,14	0,12	0,02
V_{oc} [V]	0,62	0,62	0,63	0,46
j_{sc} [mA/cm^2]	-0,98	-0,71	-0,59	-0,23
FF [%]	47,7	30,8	33,9	19,0

A concluding comparison of coating methods is not possible due the statistical deviation of individual performance data (see Figure 7.1). However, working solar cells with hybrid and polymer-fullerene photoactive layer were produced with all investigated coating methods demonstrating the potential of the set-up and process.

7.3. Polymer-fullerene solar cells

Polymer-fullerene solar cells in **standard architecture** were produced by slot-die coating on ITO-glass substrate with the batch coater and on ITO-PET substrate with the pilot coating line. Figure 7.4 shows the current density voltage curve of a P3HT-P200:PCBM device on ITO-glass substrate. The cell exhibits a good V_{OC} of 0.6 V and has a reasonable fill factor, and short current density. Though record efficiencies of up to 5 % have been achieved with P3HT-P200:PCBM (Irwin, et al. 2008), efficiencies between two and three percent are more common and can be considered state of the art for technical coating tools (see Table 1.1).

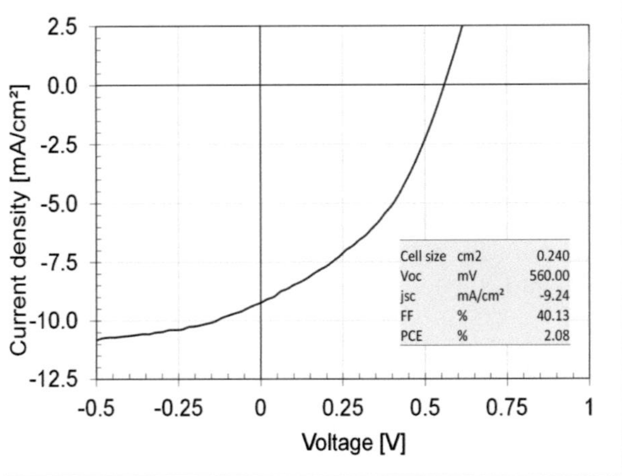

Figure 7.4: Current density voltage curve of a P3HT-P200:PCBM polymer solar cell prepared by slot-die coating on ITO-glass with the batch coater.

Layer stacks for polymer-fullerene solar cells in **inverted architecture** were realized, including fully coated charge adaptation and electrode layers. PET substrate was slot-die coated subsequently with: Ag-PR020, ZnO, P3HT-P200:PCBM, and PH1000. To realize a quick test of open circuit voltage and resistance, a silver grid was pad-printed on top. The 35 mm wide slot-die was manually displaced in cross web direction by ~2 mm after each coating to inhibit overlap between the PH1000-anode and the Ag-cathode. In this configuration a voltmeter (or solar simulator) can be attached to anode and cathode at the sides (see Figure 7.5). The area in the center where all layers overlap forms a complete solar cell stack with an aperture width of 29 mm. Figure 7.5 specifies the film thicknesses and material systems used in this experiment.

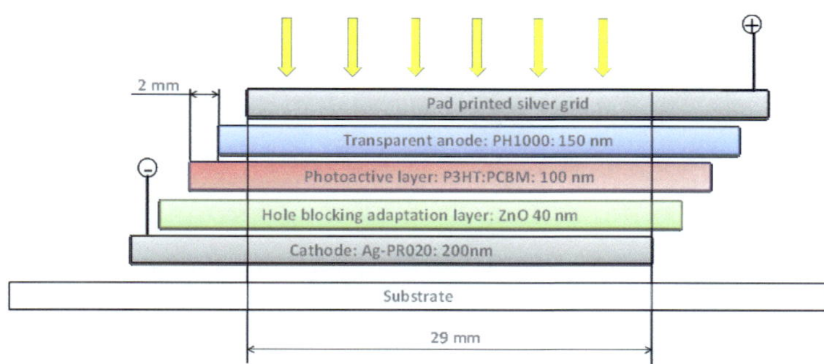

Figure 7.5: Schematic sketch of the device architecture used for inverted polymer solar cells with coated adaptation layers and electrodes.

As defects propagate in multi-layer coatings, the formation of a complete stack with good coating quality is demanding. Though coating quality was acceptable, the mechanical stability of the silver electrode was insufficient to withstand contact with the conveyor rolls. Since coating of four consecutive layers requires re-winding the film three times in the current configuration, agglomerates separated from the silver layer, creating visible pinholes. The agglomerates are then redeposit on the film and create defects in the consequent films which resulted in short currents. Possible solutions to this problem are addressed in Section 8.2. A photograph of a solar cell stack prepared by liquid phase deposition on low cost PET substrate is shown in Figure 7.6. The first four layers were slot-die coated at 6 m/min with the pilot coating line, demonstrating the potential to produce complete solar cells in a continuous multilayer coating process.

Figure 7.6: Photograph of a stack of five liquid phase processed functional films for an inverted solar cells. The first four layers were slot-die coated on the pilot coating line.

8. Summary, conclusion, and outlook

8.1. Summary

Polymer photovoltaic is a promising technology for low-cost energy conversion. Though polymer solar cells have reached power conversion efficiencies above 9 % in laboratories, the average values of large area cells are much lower. Therefore the transfer from laboratory to production scale is crucial for the success of the technology.

Design and implementation of experimental set-up

Two experimental set-ups for coating and drying of functional films for polymer solar cells were designed and employed in this work. A technical laboratory coater ("batch coater") was implemented to bridge the gap from laboratory to a pilot-scale coating line.

The batch coater was designed to enable precise control of operating conditions while at the same time employing technical tools for coating and drying. A full encapsulation allows for either cleanroom or inert operation. The scarcity of the photoactive materials, a practical obstacle for the scale-up of the coating process, was addressed by designing a miniature slot-die system with an overall hold-up of less than 2 ml. An impinging jet drier with adjustable air flow-rate and temperature, and a temperature controlled support plate realize technical coating conditions in a laboratory scale set-up. The need for a homogeneous local drying rate on a stationary substrate led to the development of measurement systems for local heat transfer coefficients and a nozzle field with local extraction of spent air.

The pilot coating line was developed in cooperation with the plant manufacturer. A modular design provided a basic platform for coating and drying while retaining the flexibility to modify the plant for a specific process. Modifications included: optimization of speed control, adaptation of the gravure coating system, implementation of a slot-die coating table, and adaptation of air filter cases and drier nozzles.

Experimental

Procedures and characterization methods are crucial for the reproducibility of experimental results. In particular, the preparation of coating ink determined whether polymers, molecules, and particles were completely dissolved or dispersed. Procedures for knife, spray, gravure, and slot-die coating were established. Methods for substrate cleaning and patterning had to be adapted from rigid to flexible substrates. Fluid dynamic properties of coating inks were determined by rotational rheology, and pendant drop measurements. Their wetting behavior was characterized by contact angle measurements of a sessile drop on typical substrates or layers. Films were characterized with respect to film thickness, topography, and roughness as well as their opto-electric properties: conductivity and light absorption. The characterization of solar cells was conducted at cooperating institutes under standardized conditions.

Material properties

Viscosity and surface tension of inks for conductive (transparent or reflective) and semi-conductive (hole-blocking, electron-blocking, or photoactive) layers were presented. Quantitative models to estimate these properties as a function of composition, shear-rate, and temperature were proposed or reviewed. The surface tension of water based PEDOT:PSS dispersions, and the specific conductivity of films were measured as function of surfactant content. At high coating speed an increase in surface energy of the substrate or underlying layer may be required to ensure wetting. The surface energy of a photoactive layer and a standard PET substrate was measured after corona treatment with variable intensity and fitted to a model as a function of web speed and corona generator power.

Coating of functional films

Polymer-fullerene and polymer-nanoparticle films and solar cells were produced by slot-die, knife, spray, and spin coating. The wet film thickness of knife coated samples was determined to be a function of coating velocity, feed volume, and material properties. A dimensionless stability window was observed for slot-die coating, and its boundary was described by a modification of the Visco-Capillary Model. For the experimental setup used in this work, spray-coated films were much rougher, and leveling occurred only at higher film thicknesses.

A comparison of surface topography and light absorption showed that the coating method had a significant impact on film morphology and properties. Spray-coated films absorbed less light at all wavelengths either due to a local variation in film thickness or due to lower crystallinity caused by rapid drying during atomization. In the lower wavelength regime, the absorption curves differ only slightly for all other coating techniques. Although spin coated films dry faster than knife- and slot-die coated films, they showed significantly higher absorption at wavelengths above 500 nm. Thus, the increase in absorption observed for spin coating could be the result of polymer alignment caused by the coating method's higher shear intensity. To verify this hypothesis, slot-die coating experiments were conducted under identical drying conditions and varying shear intensity.

Consistent with a shear-rate induced crystallization of the photoactive polymer-fullerene-blend, an increasing light absorption was observed towards higher coating speed and smaller slot width. Especially for inks composed of low boiling point solvents, a significant decrease in light absorption was measured at coating speeds below 5 m/min. Transparent conductive polymer layers showed only a small increase in specific conductivity towards lower slot-width which remained within the experimental error. However, a significant decrease was observed at higher coating velocity when the operating point approached the stability limit. Reflective silver electrodes did not exhibit a decreasing conductivity towards higher coating speeds, but an increased standard deviation of the conductivity values also indicates a decrease in film homogeneity. In contrast to the polymer fullerene solution, both dispersions (Silver nanoparticle ink and conductive-polymer nanoparticle ink) showed an increase in surface roughness towards higher coating speed, confirming a decrease in film quality.

Technical drying of functional films

In previous works, a strong impact of drying kinetics on the morphology and performance of polymer solar cells was observed under laboratory conditions. Therefore, drying experiments were conducted with the batch coater and the pilot coating line to investigate the impact of process parameters under technical conditions. A variation of air flow rate had no significant effect and an increase in drying temperature led to a moderate increase in light absorption of the photoactive layer. In contrast, a variation of solvent composition from low to high boiling point solvents resulted in a significant increase in light absorption.

Due to the low viscosity of the photoactive inks, the long drying times caused by high boiling point solvents may lead to the development of destabilizing lateral flows. In general, photoactive films become more stable at higher solid content, lower wet film thickness, and lower solvent boiling point. Since solid content and wet film thickness are typically not variable due to limited solubility and fixed wanted dry film thickness, a mixture of 80 vol.-% Toluene and 20 vol.-% Dimethyl-anisol was proposed as a compromise between optimized morphology and process stability.

Similar effects were observed for the stability of water based dispersions of high conductive polymer. An increase in drying time due to the addition of high boiling point additive Dimethyl-sulfoxide (and possibly due to higher ambient humidity) resulted in a decrease in film stability.

Solar cell devices

Solar cells composed of polymer-fullerene and hybrid photoactive layer were produced with both experimental set-ups, by all coating methods except gravure coating, on rigid (glass) or flexible (PET) substrates. Knife coated cells with 1.2% efficiency demonstrated that hybrid solar cells can be produced with large scale coating methods. Slot-die coated polymer-fullerene solar cells achieved a maximum efficiency of 2.1% (P3HT:PCBM) in a sheet to sheet process on rigid ITO-glass substrate. Finally, the potential of the technology for low cost production of solar cells was demonstrated by coating an inverted stack with liquid processed electrodes and adaptation layers. PET substrate was slot-die coated at 5 m/min with a 200 nm silver electrode, a 40 nm ZnO nanoparticle electron conducting layer, a 100 nm P3HT-P200:PCBM photoactive layer, a 200 nm PH1000 transparent electrode and a pad printed silver grid.

8.2. Conclusion

The aim of this thesis, *the investigation of technical coating and drying processes for functional films in polymer solar cells*, could only be achieved by implementing a complete process first. Many topics beside the coating and drying of films had to be addressed: e.g. the development and optimization of the experimental set-up, the definition of experimental procedures, and the measurement and modeling of material properties. In comparison to a focused view on a single aspect, this broad scope highlights the interactions between individual process steps and their implication for the overall process. As an example, an increase in coating velocity led to an increase in light absorption which was most pronounced for fast drying inks. As pilot-scale experiments are typically conducted at low coating speed (due to the high material prices) with low boiling point solvents (to reduce the required drier length or to inhibit drying related defects), this may be one reason contributing to the lower efficiencies of roll-to-roll produced solar cells in comparison to spin coated reference cells. Consequently, viscous shear can be applied intentionally in a technical process to increase light absorption of the photoactive layer.

Conventionally, coating and drying processes are designed to ensure process stability and desired film homogeneity. The quality requirements for polymer solar cells are extraordinarily high because an individual defect or inhomogeneity may propagate through the multilayer stack and cause short circuit currents or complete malfunction of the solar cell. The low film thicknesses are challenging for the coating process and the low viscosity provides little resistance against the development of destabilizing flows. It is thus necessary to limit the process parameters to operating conditions where highest homogeneity can be achieved. For the slot-die coating process this region can henceforth be described by a dimensionless coating window.

The materials used in polymer solar cells exhibit optical and electrical properties which determine device efficiency through a series of complex interactions. Since process parameters such as coating velocity or solvent composition significantly affect these properties, it is necessary to test new material in a technical coating and drying process. In many instances a tradeoff between process stability and optimization of film properties has to be made. For example, the stability of high conductive polymer coatings is increased by the addition of surfactant, whereas specific conductivity of the films drastically decreases at higher surfactant content.

An optimization of the process was demonstrated for standard material systems and the recommended process parameters can be used as a starting point for further optimization. A transfer of the general results to other material systems or set-ups is possible, but the exact values may depend on material properties, ink composition, and experimental details.

8.3. Outlook

Experimental set-up and procedure:

One of the major drawbacks of the current experimental configuration of the Pilot Coating Line is the contact of conveyor rolls to the top of the film during winding and re-winding. By using the coating cylinder of the development coater as an extension, a contactless operation could be realized with the existing equipment. In a technical production process, multilayer coating could be realized by additional coating and drying modules. A lamination unit before the up-winder could protect the cell stack from mechanical stress during winding.

One of the advantages of slot-die coating is the possibility to coat more than one layer simultaneously. A simultaneous coating of a complete cell stack seems ambitious since mixing at the layer interfaces cannot be tolerated. Instead, controlled concentration gradients could be realized by coating multiple layers with different polymer to fullerene ratios to decrease transport losses within the bulk hetero-junction layer.

Another process parameter that has not been addressed in this thesis is the process temperature. The slot-die and the substrate plate can be temperature controlled to improve coating quality or film function. An increased temperature would allow for an increase in solid content since the solubility limit increases with temperature. A higher solid content at a given dry film thickness would directly reduce the required drier length and energy to dry the film. Though the increased viscosity would reduce the highest stable coating speed, the development of instabilities and defects during drying could be reduced.

Since high shear rate during coating increases the light absorption of semi-conducting bulk hetero-junction layers, a logical step would be the design of coating tools that increases coating shear-rate. The experiments with variable slot-width could be extended to even lower slot widths or longer slots. Once sufficient experimental data is available, it may even be possible to find a quantitative correlation of coating shear intensity and film property.

The manual adjustment of coating gap and positioning of the die is a source for experimental error. Especially for intermitted coating an automated control is necessary to synchronize coating speed, die-positioning, and volume flow rate.

The capacity of the PSN-drier is sufficient for most tasks in organic electronics. For other application with higher wet film thicknesses, such as electrodes for Lithium-ion batteries, an increase in transfer coefficients and drier length may be necessary. The implementation of an additional drier segment was incorporated in the original design. The geometry of the existing setup could be improved by increasing the number of slots and by mounting nozzles onto the perforated plate.

Another improvement for the drying set-up would be the implementation of an in-line flash sinter oven. Especially for particle based films, film properties and mechanical stability could be improved significantly.

During the characterization of the drying nozzle fields used in this work, strong inhomogeneities in transfer coefficients were measured. Inhomogeneous drying can cause drying defects or inhomogeneous film properties in functional films. This effect is most critical in laboratory set-ups where the substrate remains stationary under a nozzle field, but it is also relevant for continuous web driers towards low coating speed. First steps were taken to develop a drying field with higher homogeneity. The ARN-drier applied at high nozzle to web distances showed good homogeneity though energy efficiency was low. An impinging jet nozzle field with high homogeneity and good energy efficiency has a large potential not only for organic electronic application. An advanced set-up for high definition measurement of heat transfer coefficients was developed by Cavadini who investigates new nozzle fields based on numerical simulation.

Coating of functional films

The coating method has a significant impact on film properties for photoactive films. The properties of electron or hole blocking layers are difficult to measure outside a complete solar cell. For electrode films on the other hand, the specific conductivity can be measured easily and it might be interesting to extend the investigation of coating methods to high conducting PH1000, Ag-Pr020, or other electrode materials. As well an inclusion of additional coating methods such as gravure coating or ink jet printing is possible. The spray coating process could be improved by applying more energy (e.g. by ultra sound) to the atomization of the spray.

The impact of coating shear rate could be investigated further by using even higher shear rates, other material systems, or other solvent systems. Especially the application to non-chlorinated solvents could be important for large volume production. The negative effect of "green" solvents might be compensated in part by higher shear rates.

Drying of functional films

The impact of drying rate on the stability of functional films was briefly addressed in Section 6.4. Though many problems can be solved by trial and error, the development of defects and destabilizing flows in low-viscous films during drying is far from understood. For a thorough investigation, experiments designed to investigate film stability have to be conducted. By applying property models in a drying simulation, the evolution of film properties during drying could be computed and used to define critical dimensionless numbers that predict the appearance of defects. A complete three dimensional fluid dynamic simulation of a drying film could provide additional insight.

Film properties changed at a timescale of minutes to hours during annealing. This observation has serious implications for the design of a production process, where residual solvents have to be removed prior to encapsulation. The measurement of residual solvent concentration is difficult due to the small absolute amounts in a nanometer film. A Quartz Cristal Microbalance has a theoretical resolution of a few nano-grams per centimeter squared and may be the key to identify the bottleneck for solvents desorption in nanometer films.

Validation of results and transfer to production process

The investigation of film properties was limited to specific conductivity of electrode films, light absorption spectra of photoactive films, and film topography. Methods such as grazing incident X-Ray diffraction or electron tomography may provide further insight into the structural changes induced by coating method, shear-rate or solvent composition.

Eventually, the impact of this thesis depends on whether the results can be transferred to a production process to increase the efficiency of solar cells and whether a generalization to other material systems is possible.

9. Appendix

9.1. List of publications

Patent

1. P. Cavadini, W. Schabel, P. Scharfer, **L. Wengeler**. "Device for transferring heat or mass, comprising hexagonal jet nozzles, and method for treating surface layers", in, *WO Patent* 2,013,045,026, (2013).

Publications

1. K. Peters, **L. Wengeler**, P. Scharfer, W. Schabel. "Liquid film coating of small molecule OLEDs" *Journal of Coatings Technology and Research (2013): published online 08.2013.*
2. **L. Wengeler**, R. Diehm, P. Scharfer, and W. Schabel. "Dependence of opto-electric properties of (semi-) conducting films in polymer based solar cells on viscous shear during the coating process." *Organic Electronics* (2013): 1608-1613.
3. **L. Wengeler**, K. Peters, M. Schmitt, T. Wenz, P. Scharfer, W. Schabel. "Fluid-dynamic properties and wetting behavior of coating inks for roll-to-roll production of polymer-based solar cells." *Journal of Coatings Technology and Research* (2013): Published online 04.2013.
4. **L. Wengeler**, M. Schmitt, K. Peters, P. Scharfer, and W. Schabel. "Comparison of large scale coating techniques for organic and hybrid films in polymer based solar cells." *Chemical Engineering and Processing: Process Intensification* 68 (2013): 38-44.
5. M. Schmitt, M. Baunach, **L. Wengeler**, K. Peters, P. Junges, P. Scharfer, and W. Schabel. "Slot-die processing of lithium-ion battery electrodes–Coating window characterization." *Chemical Engineering and Processing: Process Intensification* 68 (2013): 32-37.
6. **L. Wengeler**, B. Schmidt-Hansberg, K. Peters, P. Scharfer, and W. Schabel. "Investigations on knife and slot-die coating and processing of polymer nanoparticle films for hybrid polymer solar cells." *Chemical Engineering and Processing: Process Intensification* 50 (2011): 478-482.

7. **L. Wengeler**, M. Schmitt, P. Scharfer, and W. Schabel. "Designing a sensor for local heat transfer in impingement driers." *Chemical Engineering and Processing: Process Intensification* 50 (2011): 516-518.

8. **L. Wengeler**, W. Schabel. "Verfahrenstechnische Herausforderungen für die Herstellung großflächiger organischer und hybrider Solarzellen." *Nanotechnik/Photonik 4, AT-Fachverlag, Stuttgart* (2010): Published online 08.2010.

9. **L. Wengeler**, B. Schmidt-Hansberg, K. Peters, F. Rauscher, P. Scharfer, and W. Schabel. "Beschichten und Trocknen hybrider Polymersolarzellen." *Chemie Ingenieur Technik* 82 (2010): 1461-1461.

10. **L. Wengeler**, M. Schmitt, P. Scharfer, W. Schabel. "Entwicklung eines Sensors zur Messung des lokalen Wärmeübergangs in Prallstrahltrocknern." *Chemie Ingenieur Technik* 82 (2010): 1428-1429.

11. K. Peters, **L. Wengeler**, S. Hartmann, D. Bertram, P. Scharfer, and W. Schabel. "Beschichtung und Trocknung von SM-OLEDs" *Chemie Ingenieur Technik* 82 (2010): 1450-1450.

Conference contributions as first author

1. **L. Wengeler**, P. Scharfer, W. Schabel: "Large scale coating and drying processes for polymer solar cells" (Talk) *10th European Coating Symposium*, 10.-13.09.2013, Mons, Belgium.

2. **L. Wengeler**, K. Peters, P. Scharfer, W. Schabel: "Selbstdosierende Beschichtungsverfahren." (invited talk) *Hochschulkurs - Beschichtung & Trocknung von dünnen Schichten*, 14.11.2012, Karlsruhe, Germany.

3. **L.Wengeler**, M. Schmitt, K. Peters, P. Scharfer, W. Schabel: "Coating and wetting of semiconduction organic and hybrid films: Fluid-dynamic properties, process parameters and wetting behavior." (Talk) *Material Science Engineering (MSE 2012)*, 25.-27.09.2012, Darmstadt, Germany.

4. **L.Wengeler**, F. Buss, P. Scharfer, W. Schabel: "Dependence of optoelectronic properties of (semi)conducting films in polymer based solar cells on viscous shear during the coating process." (Poster) *16th International Coating Science and Technology Symposium*, 25.-27.09.2012, Atlanta, USA.

5. **L.Wengeler**, M. Schmitt, K. Peters, P. Scharfer, W. Schabel: "Coating and wetting of semi-conducting organic and hybridfilms: Fluid dynamic

properties, process parameters and wetting behavior" (Talk) *16th International Coating Science and Technology Symposium*, 25.-27.09.2012, Atlanta, USA.

6. **L. Wengeler**: "Large scale coating techniques for functional films in polymer based solar cells" (invited Talk), Kroenert open house event, 22.03.2012, Hamburg, Germany.

7. **L. Wengeler**, M. Schmitt, P. Darmstädter, C. Kroner, K. Peters, K. Köhler, P. Niyamakom, P. Scharfer and W. Schabel: "Comparison of large scale coating techniques for polymer particle films in hybrid solar cells." (Talk) *9th European Coating Symposium,* 08-10.06.2011, Abo/Turku, Finland.

8. **L. Wengeler**: "Coating, wetting and drying issues in roll-to-roll processing of organic photovoltaic." (invited Talk) *9th European Coating Symposium*, invited short course lecture, 07.06.2011, Abo/Turku, Finland.

9. **L. Wengeler**, B. Schmidt-Hansberg, F. Brunner, P. Scharfer, W. Schabel: "Investigating coating and drying of polymer and hybrid solar cells" (Poster) *International Conference on Materials for Energy (ENMAT) 04.-08.06.2010*, Karlsruhe, Germany.

10. **L. Wengeler**, M. Schmitt, P. Scharfer, W. Schabel: "Entwicklung eins Sensors zur Messung des lokalen Wärmeübergangs in Prallstrahltrocknern." (Poster) *ProcessNet 21.-23.09.2010*, Aachen, Germany.

11. **L. Wengeler**, B. Schmidt-Hansberg, K. Peters, F. Brunner, K. Döring, F. Rauscher, P. Nyamakom, P. Scharfer, W. Schabel: "Beschichten und Trocknen hybrider Solarzellen" (Talk) *ProcessNet 21.-23.09.2010*, Aachen, Germany.

12. **L. Wengeler**, B. Schmidt-Hansberg, K. Peters, P. Scharfer, W. Schabel: "Investigations on slot-die coating and processing of polymer nanoparticle films for hybrid polymer solar cells. " (Talk) *15th International Coating Science and Technology Symposium*, 13.-15.09.2010, St. Paul, USA.

13. **L. Wengeler**, M. Schmitt, P.Scharfer, Wilhelm Schabel: "Designing a sensor for local heat transfer in impingement driers." (Poster) *15th International Coating Science and Technology Symposium*, 13.-15.09.2010, St. Paul, USA.

14.**L. Wengeler:** "Beschichten und Trocknen hybrider Solarzellen." (invited Talk) Evonik Symposium and Networking Workshop, 03.2011, Marl, Germany.

15.**L. Wengeler**, B. Schmidt-Hansberg, P. Scharfer, W. Schabel, F. Rauscher, W. Hoheisel: "Processing of nanoparticle polymer mixtures for organic photovoltaic films - A project outline and first results. " (Poster) *8th European Coating Symposium*, 07.-09.09. 2009, Karlsruhe, Germany

16.**L. Wengeler**, I. de Vries, P. Scharfer , W. Schabel: "Local Heat and Mass Transfer in Slot Nozzle Driers for Organic Photovoltaic (OPV) and Organic Light Emitting Diode (OLED) Thin Films on Foil" (Poster) *8th European Coating Symposium*, 07.-09.09. 2009, Karlsruhe, Germany.

Student research projects conducted in conjunction with this thesis

1. Fabian Brunner, „Entwicklung eines Versuchsstands zur Beschichtung und Trocknung von Filmen für organische und hybride Solarzellen", Studienarbeit, Universität Karlsruhe - KIT (TH), 2010.

2. Marcel Schmitt, „Inbetriebnahme und Einstellung der Prozessparameter einer Pilotanlage zur Herstellung von organischen und hybriden Solarzellen", Diplomarbeit, Universität Karlsruhe - KIT (TH), 2010.

3. Stefan Jaiser, „Characterization of the Roll-2-Roll Impingement Dryer", Internship at Holst Centre, Eindhoven, The Netherlands, 2011.

4. Konrad Döring, „Untersuchung funktionaler Schichten für organische Leuchtdioden und Solarzellen bei unterschiedlichen Trocknungsbedingungen", Studienarbeit, Universität Karlsruhe - KIT (TH), 2011.

5. Peter Darmstädter, „Konstruktion und Inbetriebnahme eines Gravurwalzenmoduls zur Beschichtung halbleitender Filme für polymerbasierte Solarzellen", Studienarbeit, Universität Karlsruhe - KIT (TH), 2011.

6. Martin Grunert, „Aufbau eines Versuchsstandes zur Bestimmung des lokalen Wärmeübergangs in Prallstrahltrocknern", Studienarbeit, Universität Karlsruhe - KIT (TH), 2011.

7. Christian Kroner, „Vergleich unterschiedlicher Beschichtungsverfahren zur Herstellung halbleitender Polymer und Polymer-Partikel-Filme", Studienarbeit, Universität Karlsruhe - KIT (TH), 2011.

8. Christian Hotz, „Inbetriebnahme einer Batch-Anlage zur inerten Beschichtung von organische Leuchtdioden", Studienarbeit, Universität Karlsruhe - KIT (TH), 2011.

9. Thomas Wenz, „Benetzungseigenschaften von Flüssigkeiten und Oberflächen für polymerbasierte Solarzellen", Studienarbeit, Universität Karlsruhe - KIT (TH), 2012.

10. Jana Kumberg, „Thermische Behandlung von Polymer-Nanopartikel-Filmen zur Verbesserung der opto-elektrischen Eigenschaften in Polymersolarzellen", Bachelorarbeit, Universität Karlsruhe - KIT (TH), 2012.

11. Ralf Diehm, „Einfluss der Intensität viskoser Scherung auf die Eigenschaften von halbleitenden Polymer-Partikel-Filmen für polymerbasierte Solarzellen", Studienarbeit, Universität Karlsruhe - KIT (TH), 2013.

12. Harald Wiegand, „Konstuktion und Test eines mobilen Sensors zur Bestimmung der lokalen Wärmeübergangskoeffizienten in technischen Trockner", Studienarbeit, Universität Karlsruhe - KIT (TH), 2013.

9.2. List of symbols

Latin letters

a	-/m	Specific light absorption
A	-	Light absorption
A	m²	Area
B	m	Width; perpendicular to coating direction
$\tilde{c}\_P$	J/(K*mol)	Molar heat capacity
d	m	Dry film thickness
D	m	Wet film thickness
E	m/s	Rate of evaporation
E_A/R	K	Charactristic temperature for temperature dependent vioscosity
FF	%	Fill factor
g	m/s²	Gravitational acceleration
G	m	Coating gap
H_{NSD}	m	Distance from drier nozzle to substrate
I	A	Current
i_{solar}	W/m²	Incident solar power density
j	A/m²	Current density
L	m	Characteristic length
L_T	m	Nozzle spacing
m	-	Coefficient in the modified visco capillary model
n	-	Logarythmic slope of shear rate dependent viscosity
P	W	Power
PCE	%	Power conversion efficiency
PD_{Corona}	s²	Corona treatment intensity
Q	m³/s	Volume flow rate
R	Ohm	Resistance
R	Ohm	Resistance
R	m	Radius
s	-	Speed ration in gravure coating
SC	S/m	Specific film conductivity
SFE	N/m	Surface energy
SFT	N/m	Surface tension
SR	Ohm*m	Specific resistance
T	°C	Temperature
T_B	°C	Boiling point at 1 atm pressure

146

U	V	Voltage
V	V	Voltage
V_c	µm	Specific cell volume of a gravure coating roll
W	m	Slot width
$\tilde{x}$	-	Molar fraction of component i in the liquid phase
x_i	-	Mass fraction of component i
y	m	Dimension
$\tilde{y}$	-	Molar fraction of component i in the gas phase
u	m/s	Coating velocity

Greek letters

α	W/m²K	Heat transfer coefficient
β	m/s	Mass transfer coefficient
$\dot{\gamma}$	1/s	Shear rate
$\delta_{(S,A)}$	m²/s	Diffusion coefficient
$\Delta\llbracket \tilde{h}\_v \rrbracket^\wedge$	J/mol	Enthalpie of evaporation
η	Pa*s	Viscosity
θ	°	Contact angle
λ_Heat	J/(K*m)	Heat conductivity
λ_Visc	s	Inverse critical shear rate of shear rate dependent viscosity
ρ	kg/m³	Density
σ	N/m	Surface tension
τ	s	Critical time
Ω	1/s	Spin frequency

Indices

0	Zero shear
□	Per unit square of a surface
∞	Infinite shear or bulk
Crit	Operating parameter at the boundary of a stable processing window
D	Disperse
F	At the film's surface
Ful	Fullerene

i	Component
L	Liquid
Min	Minimal value
OC	Open circuit
P	Polar
Ra	Arythmetic mean roughness
Rq (RMS)	Quadratic mean roughness
S	Solid
SC	Short current

Abbreviations

ACE	Acetone
AFM	Atomic force microscope
Ag	Silver
ARN	Array of round-nozzles
ASN	Array of slot-nozzles
BC	Batch coater
BHJ	Bulk heterojunction
CB	Chlorobenzene
CF	Chloroform
DC	Development coater
DCB	Dichlorobenzene
DMSO	Dimethylsulfoxid
HEPA	High particulate efficiency air filter
HOMO	Highest occupied molecular orbital
HSC	Hybrid solar cells
HTC	Heat transfer coefficient
IND	Indane
IPA	Isopropanol
ITO	Indium tin oxide
ITO-Glass	Indium tin oxide coated glass
KNC	Knife coating
LUMO	Lowest unoccupied molecular orbital
MD	Machine direction
MeOH	Methanol
MPP	maximum power point
OPV	Organic photovoltaics
P3HT	poly(3-hexylthiophene)
PAA	Polyacrylamide
PC	Pilot coating line
PCBM	Phenyl-C61-butyric acid methyl ester
PEDOT:PSS	Poly(3,4-ethylenedioxythiophene) poly(styrenesulfonate)
PET	Polyethylenterephthalat
PS	Polystyrene
PSC	Polymer solar cell
PSN	Array of slot nozzles in the drier of the pilot coating line
PYR	Pyridin
QD	Quantum dots
R2R	Roll to roll

RMS	Rout mean square
SDC	Slot-die coating
SNC	Spin coating
SRC	Spray coating
TD	Transferse to machine direction
TOL	Toluene
VCM	Visco Capillary Model
XYL	Xylene
ZnO	Zinc oxide

Dimensionless groups

Π_{SFE}	$\dfrac{\sigma_{Solid}-\sigma_{Solid,Min}}{\sigma_{Solid,Max}-\sigma_{Solid,Min}}$	Dimensionless increase in surface energy
Ca	$\dfrac{\eta u}{\sigma}$	Capillary Number
G*	$\dfrac{G}{D}$	Dimensionless coating gap
Le	$\dfrac{\lambda_{Heat}}{\tilde{\rho}\tilde{c}_p \delta_{i,Air}}$	Lewis Number
Ng	$\dfrac{\rho \cdot g \cdot L_{Slot}}{\Delta p_{Slot}}$	Gravitational Number
Nm	$\dfrac{\rho \cdot \dot{V}^2}{\Delta p_{Slot} \cdot A_{Cavity}}$	Momentum Number
Nu	$\dfrac{\alpha L_c}{\lambda}$	Nusselt Number
Nv	$\dfrac{\Delta p_{Cavity}}{\Delta p_{Slot}}$	Viscous Number
Pr	$\dfrac{\eta c_P}{\lambda_{Heat}}$	Prandtl Number
Re	$\dfrac{\rho u D}{\eta}$	Reynolds Number
Sc	$\dfrac{\eta}{\rho \delta_{i,Air}}$	Schmitt Number
Φ	$\dfrac{D}{V_C}$	Transfer coefficient in gravure coating

9.3. Technical documentation

9.3.1. Design of coating die cavities

The function of a coating die is to distribute the coating fluid in cross web direction. To achieve a constant flow rate, the pressure drop from slot inlet to every point in cross web direction at the die-lip has to be equal. Design concepts include a tapered cavity design "coat hanger", a diverging channel design "FMP-slot-die", or an infinite cavity design. In this work an infinite cavity design was selected, because it is less specialized to a specific viscosity compared to the coat hanger and easier to clean compared to the FMP-slot-die. A sketch of a cross section of a single cavity slot-die (a) and a double cavity slot-die (b) is shown in Figure 9.1.

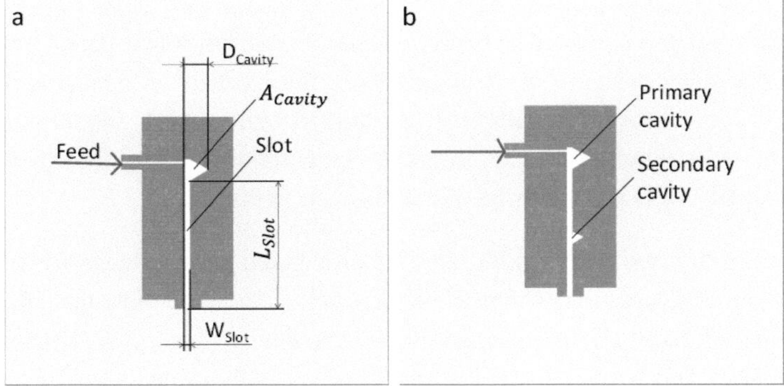

Figure 9.1: Schematic sketch of the cross section of an infinite cavity slot-die with one (a) or two (b) cavities. The most important geometrical dimensions are the length L and with (W_{Slot}) of the slot, and the depth D_{Cavity} and cross sectional area A_{Cavity} of the cavity.

The design of a coating die cavity includes the definition of geometry and size of the cavity, and the length and width of the slot for a desired wet film thickness, coating width, and web speed. The expected range of operating points and material properties have to be defined first. Table 9.1 lists typical values for the most important variables.

Table 9.1: List of expected values for operating conditions and material properties

Expected operating points	min	typical	max
Wet film thickness [μm]	1	10	100
Coating width [mm]		35 or 50	
Coating speed [m/min]	0.2	3	20
Material properties			
Viscosity [mPas]	1	4	10

The fluid dynamic flow through a three dimensional cavity can be simulated by computational fluid dynamics (CFD). However, it is difficult to evaluate the effect of non-ideal geometries e.g. the mechanical precision of the slot width in a CFD simulation. Even for ideal geometries, the large number of independent variables increase computational cost and prevent an exclusively CFD based design of coating die cavities.

In an infinite cavity design, the fluid is distributed perpendicular to coating direction due to viscous forces if the pressure drop in the die's slot (Δp_{Slot}) is much higher than the pressure drop in the cavity (Δp_{Cavity}) (Kistler and Schweizer 1997). To ensure that the flow is dominated by viscous, rather than inertia or gravitational forces, the Momentum Number (Nm) and Gravity Number (Ng) have to be small:

$$Nm = \frac{\rho \cdot \dot{V}^2}{\Delta p_{Slot} \cdot A_{Cavity}} \ll 1 \; ; \; Ng = \frac{\rho \cdot g \cdot L_{Slot}}{\Delta p_{Slot}} \ll 1; \qquad (9.1)$$

with $\dot{V}$ being the volume flow rate, ρ the fluid density, A_{Cavity} the cross sectional area of the cavity, g the gravitational acceleration, and L_{Slot} the length of the slot. For practical values, this is normally true if $\Delta p_{Slot} > 10 \, mbar$.

Assuming a laminar flow and ideal distribution within the slot, the pressure drop over the slot can be estimated:

$$\Delta p_{Slot} = \frac{12\dot{V}^2 \cdot L_{Slot} \cdot \eta_{slot}}{W_{Slot}^3}, \tag{9.2}$$

where η_{Slot} is the viscosity in the slot and W_{Slot} its geometrical depth (Kistler and Schweizer 1997).

The pressure drop in the cavity can be computed from the Viscous Number Nv which can be expressed as:

$$Nv = \frac{\Delta p_{Cavity}}{\Delta p_{Slot}} = \frac{2B \cdot \eta_{Cavity} \cdot \dot{\gamma}_{Cavity}}{D_H \Delta p_{Slot}}, \text{ with } \dot{\gamma}_{Cavity} = \frac{4\dot{V}}{A_{Cavity} \cdot D_H}, \tag{9.3}$$

assuming a laminar flow through the cavity with hydrodynamic diameter D_H and cross section of the cavity A_{Cavity} (Kistler and Schweizer 1997). The coating homogeneity in cross web direction (x) can be estimated by solving Equation (9.4) for the volumetric flow rate and calculating its relative differential:

$$\frac{d\dot{V}}{\dot{V}} = \frac{\Delta p_{Cavity}(x)}{\Delta p_{Slot}} + \frac{3dW(x)}{W_{Slot}} - \frac{dL(x)}{L} - \frac{d\eta(x)}{\eta}. \tag{9.4}$$

The length of the slot is constant in an infinite cavity design and viscosity is assumed to be Newtonian to yield an approximation for the inhomogeneity in volume flow:

$$\frac{\Delta \dot{V}}{\dot{V}} = \frac{\Delta p_{Cavity}}{\Delta p_{Slot}} + \frac{3\Delta W_{SLot}}{W_{Slot}}. \tag{9.5}$$

For a single-cavity coating die it is now possible to compute $\frac{\Delta \dot{V}}{\dot{V}}$ as a function of W_{Slot}, L_{Slot}, D_{Cavity}, ΔW_{Slot}, $\dot{V}$, B, D_H, A_{Cavity}, and $\eta(\dot{\gamma})$. In a technical die, the difference in slot width ΔW_{Slot} is a finite value, determined by manufacturing precision. Since the shape of the cavity is often defined by manufacturing method (e.g. triangular or semicircular), coating width, viscosity, and volume flow-rate are dictated by the process, the remaining design variables are L_{Slot}, W_{Slot}, and D_{Cavity}.

For a double-cavity slot-die (Figure 9.1.b), the volume flow that needs to be distributed in the second cavity is unknown. Equation (9.6) can be used as a rough approximation.

$$\dot{V}_{Cavity2} = \left(\frac{\Delta\dot{V}}{\dot{V}}\right)_{Cavity1} \cdot \dot{V} \; . \tag{9.6}$$

In analogy to a single cavity slot-die, there are three independent design parameters for the second cavity and slot. The relative deviation in volume flow can now be calculated for different values of the design variables. As an example, Figure 9.2 shows the relative deviation in volume flow for different primary slot widths and primary cavity depths.

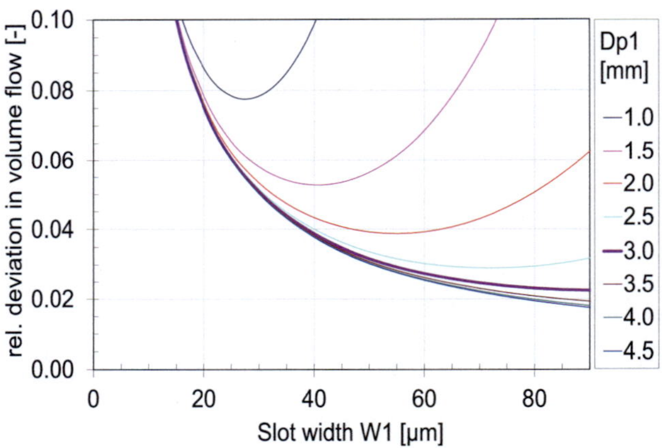

Figure 9.2: Approximation for the relative deviation of volume flow as a function of primary slot width for different primary cavity depths. Equation (9.8), (9.9), (9.10), and a constant slot length $(L_1+L_2)=7cm$ were used in this calculation.

Additional constrains and design rules may be applied and are listed below:

$$W_{Slot,1} = W_{Slot,2} \sim 2D \text{ (for shim based design)} \tag{9.7}$$

$$\frac{W_{Slot,1}}{W_{Slot,2}} = \frac{3}{5} \text{ (for variable slot width design)} \tag{9.8}$$

$$\frac{L_{Slot,1}}{L_{Slot,1}+L_{Slot,2}} = \frac{2}{3} , \tag{9.9}$$

$$\frac{D_{Cavity,1}}{D_{Cavity,2}} = 2 , \tag{9.10}$$

$$\Delta p_{Slot,1} + \Delta p_{Slot,2} > 10 \; mbar, \tag{9.11}$$

$$V_{holdup} < 1 \; ml, \tag{9.12}$$

$$\left(\frac{\Delta \dot{V}}{\dot{V}}\right)_{Cavity2} < 5 \; \%. \tag{9.13}$$

The formulas presented in this section present a possibility to approximate the effect of geometrical die designs on pressure drop, Viscous Number, and relative deviation of volume flow in cross web direction. A promising configuration can then be simulated by computational fluid dynamics or tested experimentally.

9.3.2. Lists of materials

Table 9.2: List of materials for the mini-slot-die coater

Pos	Bezeichnung	Zeichnungs NR	Werkstoff	Bemerkungen	Stck.
1	Düsenplatte1	2009-10-BaCo-1	VA	Vorderseite und Lippe geschliffen	1
2	Düsenplatte1	2009-10-BaCo-2	VA	Vorderseite und Lippe geschliffen	1
3	Einstellwinkel	2009-10-BaCo-3	AL		1
4	Kufen	2009-10-BaCo-4	AL		2
5	Schlauchanschluss	2009-10-BaCo-5	nicht rostend		1
6	Selbstklebende Teflondichtung		Teflon		
7	Abstands folien		1.4310 CrNi		
8	O-Ring		Viton	4X1	5
9	Mikrometerschraube			Einbaumessschraube 0-13 mm	2
10	Passstift		VA	5h6	2
11	Innensechskant M4X10		VA	M4X10	4
12	Innensechskant M5X30		VA	M5X30	6
13	Innensechskant M5X20		VA	M5X20	2

Table 9.3: List of materials for knife chamber for gravure coating.

Pos	Bezeichnung	Zeichnungs NR	Werkstoff	Bemerkungen	Stck.
1	Tauchwanne	2011-01-KaRa-1	Alu		1
2	Blende	2011-01-KaRa-2	Alu/VA		2
3	Bodenplatrte	2011-01-KaRa-3	Alu		1
4	Querstrebe	2011-01-KaRa-4	Alu		1
5	T-Profil	2011-01-KaRa-5	Alu		1
6	Auflagerblech	2011-01.KaRa-6	Alu/VA		2
7	Dichtungen Seite		Teflon		2
8	Rakelmesser		VA		2
9	Mikrometerschrauben			MW Import	3

Table 9.4: List of materials for the slot-die coating table in the pilot line.

Pos	Bezeichnung	Zeichnungs NR	Werkstoff	Bemerkungen	Stck.
A1	Arm 1	2010-06-SGB-11	Alu	173x140x20	1
A2	Arm 2	2010-06-SGB-11	Alu	173x140x20	1
A3	Arm 3	2010-06-SGB-07	Alu	115x35x14	1
A4	Arm 4	2010-06-SGB-07	Alu	115x35x14	1
A5	Arm 5	2010-06-SGB-05	Alu	285x80x20	2
B1	Balken	2010-06-SGB-02	Alu	90x30x20	2
H1	Hülse 1	2010-06-SGB-10	Alu	110x14x13	2
H2	Hülse 2	2010-06-SGB-10	Alu	111x14x3	2
H3	Hülse 3	2010-06-SGB-09	Alu	30x20x10	2
P1	Platte 1	2010-06-SGB-06	Alu	578x120x20	1
P2	Platte 2	2010-06-SGB-10	Alu	55x20x10	5
P3	Platte 3	2010-06-SGB-03	Alu	618x100x20	1
P4	Platte 4	2010-06-SGB-04	Alu	159x100x10	2
P5	Platte 5	2010-06-SGB-09	Alu	300x258x10	1
T1	Traverse 1	2010-06-SGB-08	Alu	618x30x20	1
W1	Winkel 1	2010-06-SGB-01	Alu	110x47x30	1

9.3.3. Technical drawings and specifications

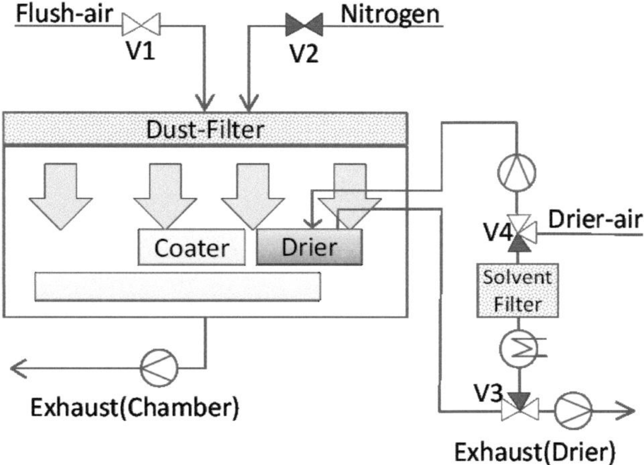

Figure 9.3: Schematic flow chart of the batch coating system in the original version.

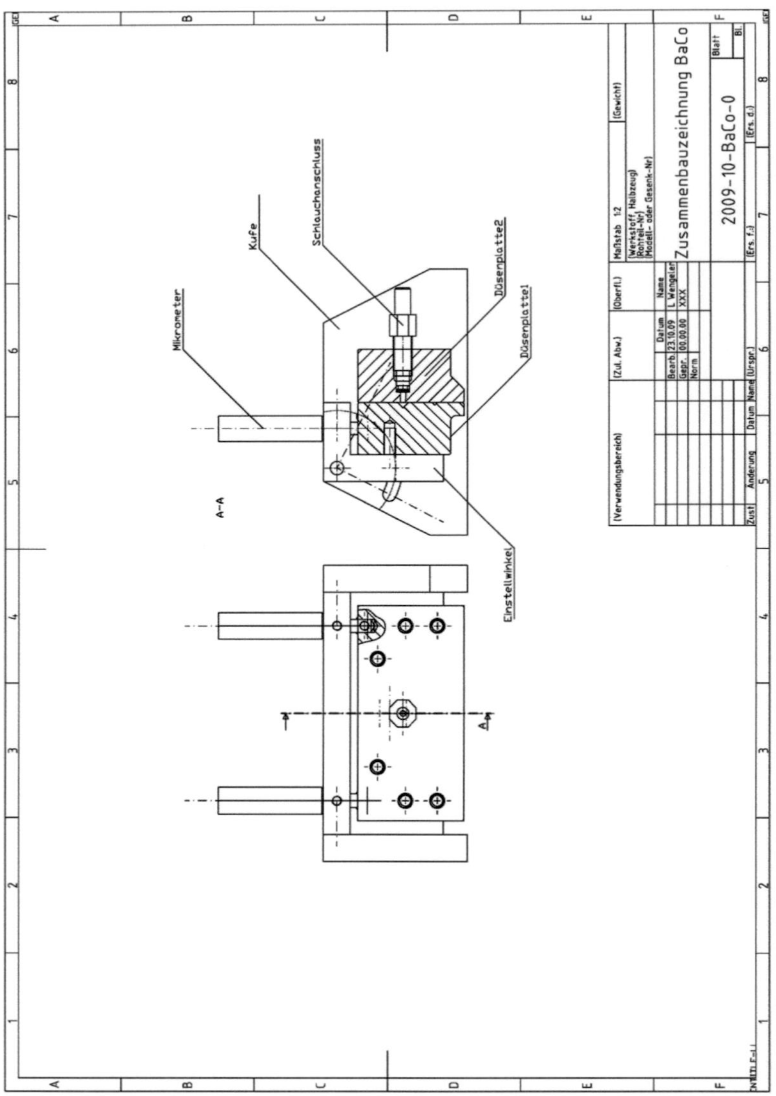

Figure 9.4: Assembly drawing of the 1st generation slot-die system. The overall hold-up (including pumps and tubes was reduced to less than 2 ml.

A runner system allowed for precise positioning and compatibility to Batch-Coater and Pilot-Coater.

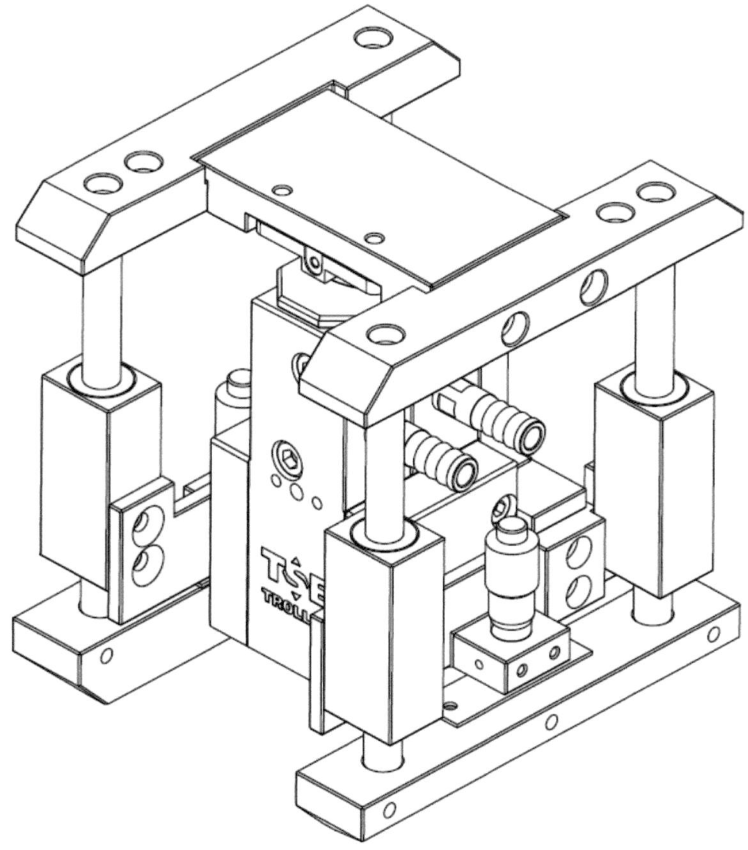

Figure 9.5: Isometric view of the 2nd generation slot-die system. Linear roll bearings are used as guide, only. Positioning is done by micrometer calipers pressing against the support frame with constant pneumatic pressure.

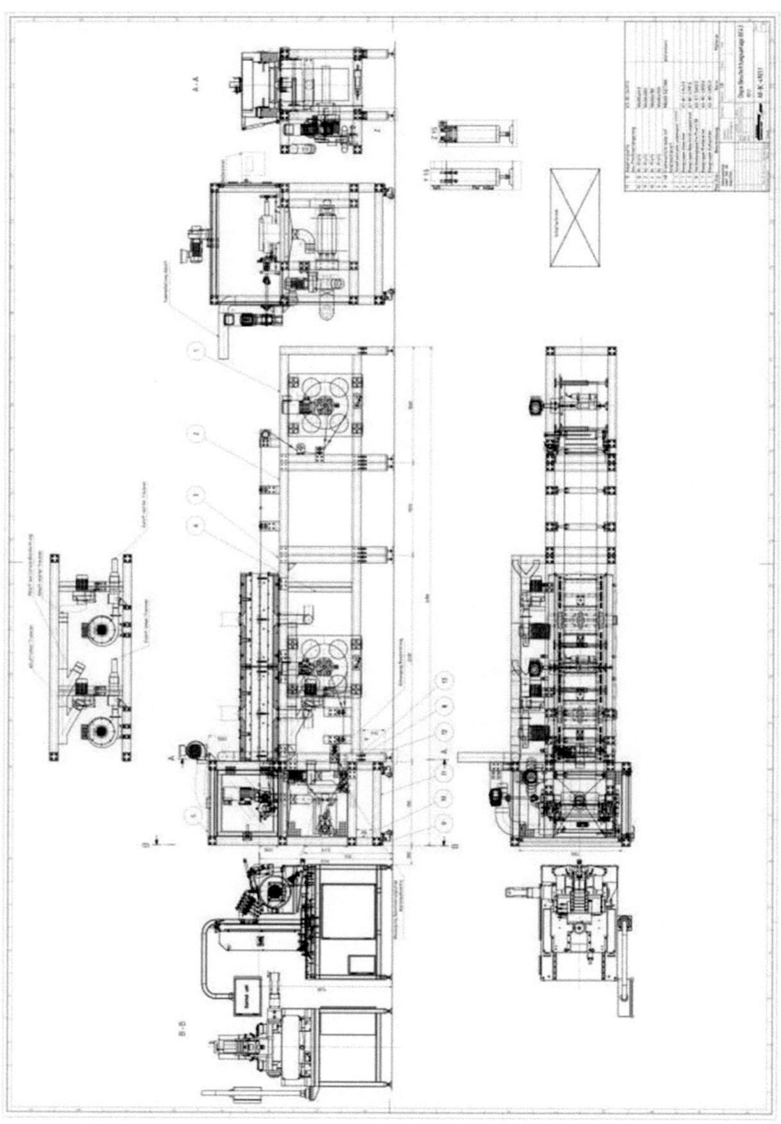

Figure 9.6: Overall assembly drawing of the pilot-scale coating line.

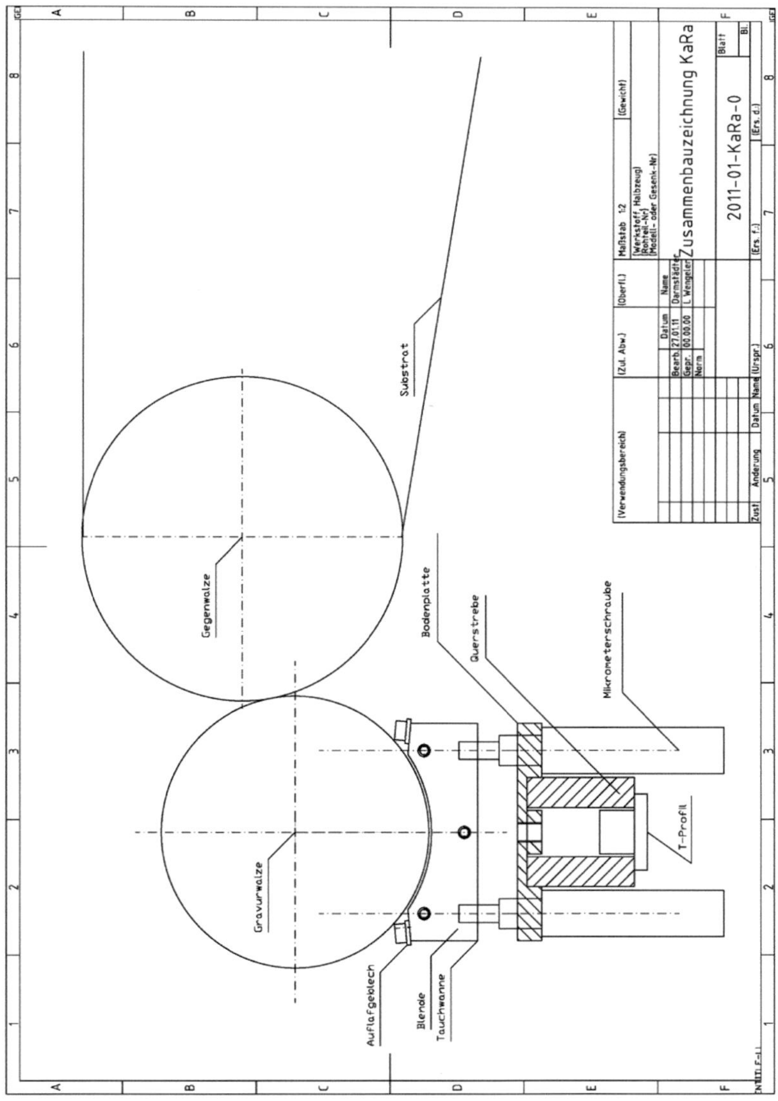

Figure 9.7: Assembly drawing of the closed chamber system for gravure coating. Hold-up was reduced to ~3 ml and evaporation was inhibited by a

sealed reservoir. The micrometer calipers can be exchanged with pneumatic cylinders.

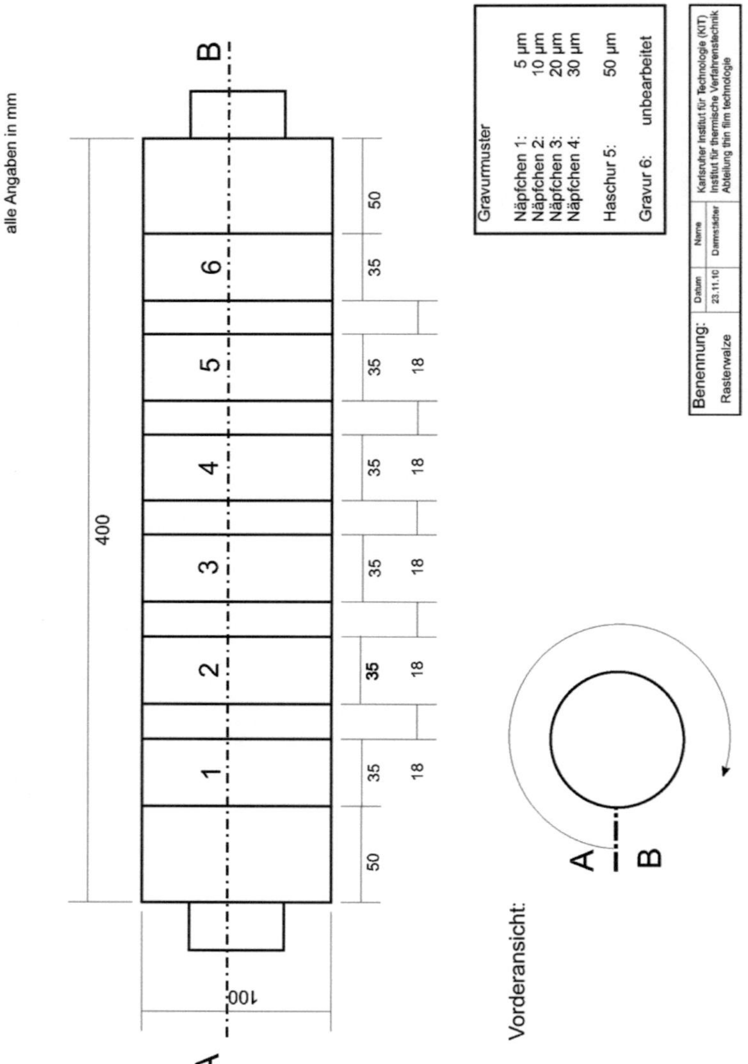

Figure 9.8: Drawing of the gravure coating cylinder detailing the positions of the 5 gravure ribbons.

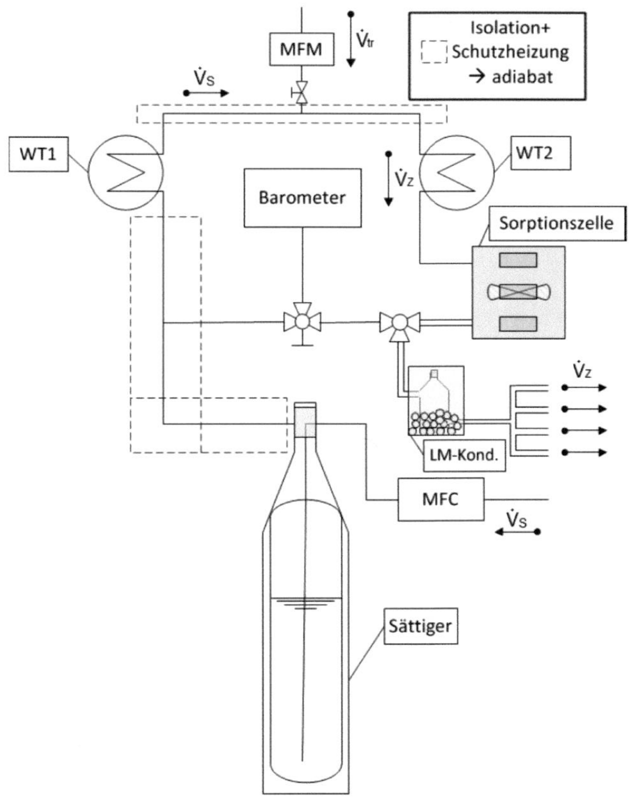

Figure 9.9: Flow-chart of the solvent annealing set-up. Flow rate controlled saturated and dry air are mixed and feed to a sorption chamber. All tubes and elements are heated in order to inhibit condensation.

9.4. Measurement of local heat transfer coefficients

In technical dryers, films and coatings are conventionally dried by impinging jets using conditioned gas. The drying process can be described by taking into account the diffusion within the film, the phase equilibrium at the film surface, and the convective mass transport in the gas phase (Schabel and Martin 2010). Phase equilibrium and diffusion within the film relate to material properties that can be measured in laboratory experiments. The heat and mass transfer in the gas phase can be described by transfer coefficients. It is possible to determine either mass or heat transfer coefficients and relate the value of the other through the Lewis correlation (Equation (6.4)). The dimensionless heat transfer coefficient or Nusselt-Number is defined as:

$$Nu = \frac{\alpha L_c}{\lambda}, \qquad Nu = f(Re, Pr, \frac{D}{H}, ...) \tag{9.14}$$

where α is the heat transfer coefficient, L_c is a characteristic length and λ is the thermal conductivity of the boundary layer. Empiric correlations can be used to approximate the Nusselt Number of a technical drier as a function of air flow rate (~Re Number), air properties (~Pr Number) and geometric dimensions. However, technical driers often vary from standard geometries making a precise dimensioning or scale-up impossible. Especially when drying films for organic photovoltaic or light emitting diodes (OLEDs) it is important to achieve a homogeneous drying rate in cross web direction. This requires local values for the transfer coefficients which are not provided by most correlations. Two approaches to access this quantity are simulation and measurement. Computational fluid dynamic simulations of existing driers are difficult because geometries are often complex and irregular. A numerical simulation of a complete drier, operating at high jet velocities, would cause high computational costs. As well the boundary conditions are not always well defined and assumptions and simplifications have to be made. Most measurement techniques measure either mass or heat transfer of a certain area and are often limited to integral values. The measurement of mass transfer typically involves large experimental efforts and may lead to a contamination of the set-up with foreign substances.

In this section, two methods are described for the measurement of local heat transfer coefficients. The first subsection (9.4.1) describes the design of a mobile sensor that can be easily transported and used to determine transfer coefficients in existing technical driers. The second subsection describes the first version of a stationary test rig to characterize nozzle geometries, featuring higher local definition and higher precision compared to the mobile sensor (Subsection 9.4.2).

9.4.1. A mobile sensor for technical driers

The major challenge for the design of a mobile sensor was the small device thickness necessary to minimize the disturbance of the flow pattern within the drier. In general it is possible to measure either heat uptake or heat loss to determine the heat transfer coefficient. But since the thickness of the device should be as small as possible, the measurement of heat loss is favorable because it is easier to introduce energy than to extract it from a small volume. The figures in this subsection were taken from the Studienarbeit of Harald Wiegand, which provides further details on the development of the sensor.

Concept

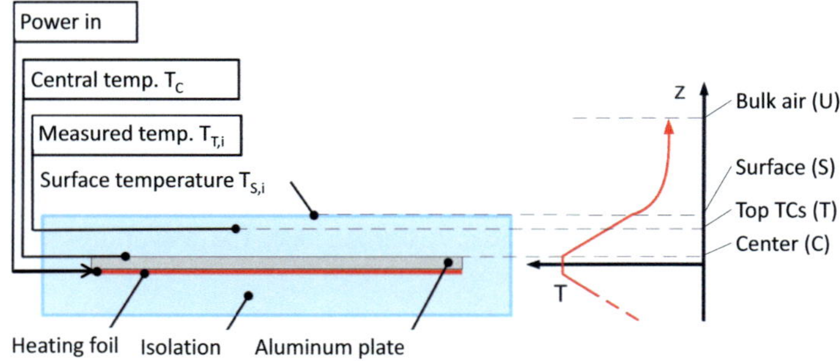

Figure 9.10: Sketch of the sensors cross section (left) and temperature profile (right).

The concept employed here is illustrated in Figure 9.10. The heat loss of a flat sensor with a planar heat source in its center is measured during impingement of cold air. Once steady state is reached, the temperature drop over the isolation $(T_C - T_{T,i})$ is measured to determine the heat flux at a certain position "i".

$$\dot{q}_i'' = \left(T_C - T_{T,i}\right) \cdot k_{1,i} \qquad (9.15)$$

The local surface temperature $(T_{S,i})$ and the temperature of surrounding air (T_U) can then be used to calculate the corresponding local heat transfer coefficient at position "i":

$$\alpha_i = \frac{\dot{q}''}{T_{2,i}-T_U}. \qquad (9.16)$$

Manufacture

Different materials were tested for the isolation and the cover layer. The best results were achieved with polyester resin (Figure 9.11). A box form was used to cast a 5 mm thick base layer of polyester resin (B; 8-49433/ARTIDEE Polyesterglas-Gießharz, Gerstäcker). Screw nuts (A) with sealed inner winding can be positioned on the bottom and can later be used for the mounting of the sensor. A 160x160 mm² heating foil (C; 189216-U1/Heizfolie, 12V, 6W, Conrad Elektronik) was glued onto a 2mm Aluminum plate (D) and positioned on top of the base layer. Two thin wire thermocouples (E; SA1-T1-3M, Type K, Omega) are glued to the aluminum and a 5 mm isolating layer is casted on top (F). Thin wire thermocouples (G) are positioned on top of the isolating layer and covered with a final thin layer of polyester resin (H). The whole sensor was then stored for >48 hours at elevated temperature (60°C). Due to volume shrinkage, the edges were slightly elevated. To achieve a flat surface, the elevations were removed by grinding.

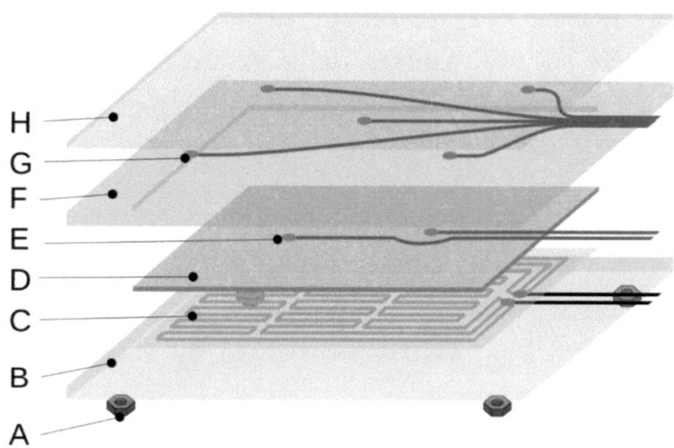

Figure 9.11: Exploded assembly drawing of the 5 point sensor for local heat transfer coefficient. All elements were merged in a casting process using polyester resin.

Calibration

Since the thermocouples are covered by a thin layer of epoxy resin the temperature at the surface of that coating $(T_{S,i})$ has to be estimated first. As all other temperatures can be measured directly, Equation (9.17) can be solved for $T_{S,i}$.

$$\dot{q}_i'' = \left(T_C - T_{T,i}\right) \cdot k_{CT,i} = \left(T_{T,i} - T_{S,i}\right) \cdot k_{TS,i}. \tag{9.17}$$

The variables $k_{CT,i}$ and $k_{TS,i}$ represent heat transport resistances through the isolating layer from the center level to the thermocouple level, and from the thermocouple level to the surface level, respectively. These values depend on thickness and thermal conductivity of the epoxy film and were determined in two experiments.

The sensor was positioned between two temperature controlled plates and isolated to the surrounding. Heat conducting paste was used to ensure that there was no significant temperature drop from the heating plates to the sensor's surfaces.

In the **first experiment**, the heating foil was not connected and a homogeneous heat flow was enforced by applying different temperatures to the top and bottom of the sensor. In the **second experiment**, the temperature controlled plates were set to a constant temperature and the heating foil was connected to a power supply. The absolute value of heat introduced in this set-up was assumed to be equal to the electric power of the heating foil.

In principle only one experiment is necessary to determine the values for all $k_{CT,i}$ and $k_{TS,i}$ values. To increase experimental certainty, several experiments with different temperature gradients and electric power were conducted and $k_{CT,i}$ and $k_{TS,i}$ values were fitted by minimizing the sum of square errors in all experiments.

Measurement procedure and discussion

To perform a measurement, the sensor is placed into a drier and a voltage is applied to the heating foil. For optimal results the value of the applied voltage is adjusted so that the temperature drop within the sensor $(T_C - T_{T,i})$ is similar to the temperature drop in the boundary layer $(T_{S,i} - T_U)$. The temperature can be monitored over time and once in steady state (typically ~45 min), the temperatures are recorded and heat transfer coefficients are computed.

At moderate inhomogeneity in heat transfer coefficients $((\alpha_{Max} - \alpha_{Min}) <$ $<100W/m^2K)$ the statistical deviation of the measured value was observed to be approximately 10%. In this regime, the magnitude was consistent with values predicted by Nusselt-correlations (see e.g. Figure 9.17).

At high inhomogeneity in heat transfer coefficients $((\alpha_{Max} - \alpha_{Min}) \gg 100W/m^2)$, a significant discrepancy to correlations and literature values was observed. Even negative values for the heat transfer coefficient were measured when trying to reproduce a literature value of an individual impinging jet (Khan, et al. 1982.).

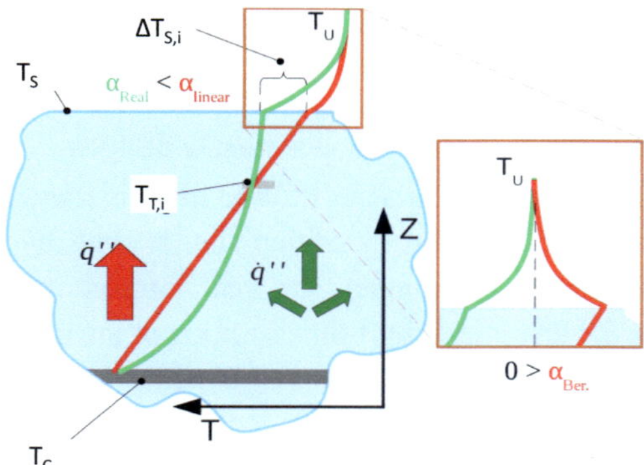

Figure 9.12: Sketch of temperature profiles during measurement of linear (red) and nonlinear (green) temperature profiles. In the case of strong inhomogeneity of heat transfer coefficients, lateral heat conduction becomes relevant and the assumption of a linear temperature profile is no longer valid.

Kahn specifies a heat transfer coefficient of 1000 W/m²K at the center of the jet and a steep decrease to less than 100 W/m²K at a radius of 60 mm from the center. A significant amount of heat is conducted in lateral direction and a nonlinear temperature profile develops within the isolation. Since the calculation of heat flux is based on the assumption of a linear temperature profile, the heat transfer coefficient can no longer be calculated. The apparent surface temperature is lower than the real value (see Figure 9.12). In some cases it can even be less than the ambient temperature resulting in negative values for the heat transfer coefficient (see insert in Figure 9.12).

The data acquisition and calculation of heat transfer coefficients was automated with a NI LabVIEW routine. The whole setup, consisting of the sensor, an electric power source, and a USB-Thermocouple DAQ module, can be easily transported. The disadvantages of this set-up are the long time required to reach steady state, the limitation to moderate difference in heat transfer coefficient, and the low spatial resolution (determined by the number of top thermocouples). Higher differences in heat transfer and a higher spatial resolution were realized in a high resolution set-up for heat transfer coefficient measurements.

9.4.2. A high resolution set-up for new nozzle designs

If the surface temperature of a semi-infinite plate is known during a step change in convective boundary condition (i.e. heating by impinging jets from isothermal initial conditions), the heat transfer coefficient can be evaluated using the analytical short time solution of the transient heat transfer. The development of this set-up was conducted in cooperation with Cavadini, who continued working on the topic and has by now developed a set-up that features pneumatic displacement of the measurement plate, improved temperature control, improved calibration (set-up and method), and automated data acquisition and analysis (Cavadini, et al. 2013b). The figures in this subsection were taken from the Studienarbeit of Martin Grunert, which provides further details on the development of the set-up.

Concept

In this subsection a set-up is described that was used for a qualitative characterization of ARN-drier (see Subsection 9.5.1). The temperature distribution on the surface of a target plate during the first seconds of heating provides qualitative information about the homogeneity of the local heat transfer. The surface temperature of a transparent PMMA plate under an array of nozzles was visualized using thermochromatic liquid crystals (TLCs) on the surface of the target plate (see Figure 9.13).

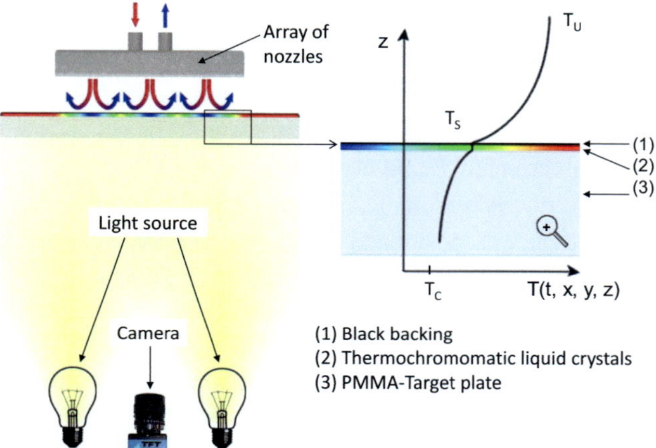

Figure 9.13: Schematic sketch of the set-up for high resolution measurement of local heat transfer. The surface temperature of a transparent PMMA plate under an array of nozzles is visualized by a layer of thermochromatic liquid crystals and recorded with a camera from below.

TLCs change the wavelength reflected from incident light with temperature, hence they can be used to capture temperature information using a CCD camera.

Manufacture

The 50 mm transparent PMMA (Plexiglas plate) was spray coated subsequently with TCs (Sprayable chiral nematic coating SPN IR 30C5W LOT# 5714, LCR Hallcrest LTD) and black backing (SPBB LOT# 110 112-1, LCR Hallcrest LTD). The plate was mounted in the focal plane of a digital camera (DFK-21BUC03, The Imaging Source Europe GmbH). To inhibit reflections, the set-up was enclosed with black cloth and illuminated by LEDs (Cree LED, XREWHT-L1-7A-Q2, LED-Tech.de). The light was directed by mirrors in a flat incidence angle onto the PMMA plate so that direct reflections of the light source were outside the angle of view of the camera.

Calibration

The camera records RGB color values for every pixel as a function of time. To correlate a pixels color to the surface temperature, the RGB values were first converted to HSV color space. In contrast to RGB values, the H-value (Hue) of thermochromatic liquid crystals is a strictly monotonous function of temperature. For calibration, TLCs were coated onto a temperature controlled plate which was positioned within the experimental set-up under identical illumination and camera position as the target plate.

Measurement procedure and discussion

To determine the homogeneity of a nozzle field, the drier heater and blower were set to a constant temperature and volume flow. Temperatures of 39°C - 60°C were used depending on the volume flow and nozzle to plate distance. Once the temperature and volume flow were at steady state, the nozzle field was manually positioned on the PMMA plate with an isolating cover in-between. The camera was started and the isolating cover was removed simultaneously.

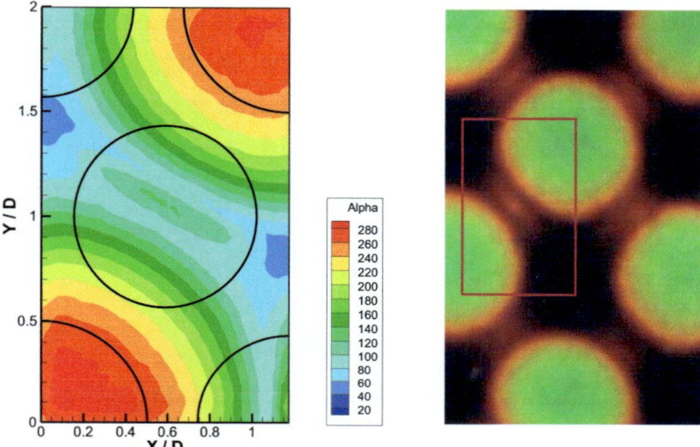

Figure 9.14: Comparison of heat transfer coefficient values taken from a numerical simulation (left, taken from (Schlegel 2011)) to a measured temperature distribution (right). The section of the nozzle field that was simulated is indicated in the picture by a red rectangle.

A measurement after 6 sec (right) is compared to a CFD simulation[14] of the heat transfer coefficient (left) in Figure 9.15. The positions of the nozzles are indicated by black rings in the simulation and the excerpt of the nozzle field that was simulated is indicated by a red rectangle in the measured image. The higher heat transfer directly under the outlet nozzles, causes higher surface temperatures and even a secondary maximum is visible in simulation and measurement.

9.5. Characterization of drying set-up

Functional films for polymer solar cells need to be dried under precisely controlled conditions in order to allow for a simple scale-up. In this work three different drier systems were used. An **array of round-nozzles drier (ARN-drier)** was used to dry films in the technical laboratory coater (see Subsection 2.2.3). The system was designed to achieve homogeneous drying without relative motion between substrate and drier. To realize even

---

[14] The simulation was made by Ute Schlegel in her Diploma thesis Schlegel, U. (2011). Numerische Untersuchungen des Stoffübergangs in Prallstrahlsystemen bei simultaner Absaugung. Institut für Thermische Verfahrenstechnik, Karlsruher Institut für Technologie..

higher heat transfer coefficients, an **array of slot nozzle drier (ASN-drier)** was implemented into the Batch-Coater. The slot-nozzles design is directly transferable to roll-to-roll systems but requires a periodic back and forth movement of the substrate in order to emulate a continuous process. The basic **pilot-line slot-nozzle drier (PSN-drier)** was not optimized in this work but was characterized so that results can be reproduced and trans-ferred to other systems.

9.5.1. Array of round-nozzles drier (ARN-drier)

The design of the ASN-drier was conducted according to the procedure proposed in Chapter Gk of the VDI-Wärmeatlas (Martin 2002). As a start-ing point, the nozzle to substrate distance (H_{NSD}) was set to 20 mm. The op-timal nozzle diameter (D_{Nozzle}) and spacing (L_T) are then defined as:

$$D_{Nozzle} = \frac{1}{5} H_{NSD} = 4\ mm, \tag{9.18}$$

and

$$L_T = \frac{7}{5} H_{NSD} = 28\ mm, \tag{9.19}$$

according to the Gk VDI-Wärmeatlas (Martin 2002). The concept of alter-nating inlet and outlet nozzles (see Subsection 2.2.3) was realized by a stack of stainless steel plates (Figure 9.15).

Figure 9.15: Photo of the compact array of round nozzles.

The compact dimensions and solvent resistance of the material are important for the application in the enclosure of the Batch-Coater.

The distribution of inlet and outlet air is illustrated in Figure 9.16. The inlet air (orange) is led through the first three plates (left to right, Figure 9.16) into a distributing cavity and from here through round nozzles towards the substrate. The outlet air is distributed to the corners by channels in second plate. It bypasses the inlet cavity in the corners and is distributed by an alternating arrangement of perpendicular channels in the 7th to 8th plate.

Figure 9.16: Exploded assembly drawing of the ARN-drier. Inlet and outlet air channels are highlighted by orange and green color, respectively.

The plates can be easily manufactured by laser cutting and the arrangement can be adapted to other flow rates or nozzle designs by adding or exchanging plates.

Integral heat transfer coefficient

The heat transfer coefficient α can be approximated using the correlations for ARN. The Nusselt number (Nu) is described as function of Reynolds (Re) and Prandl number (Pr):

$$Nu_{ARN} = \frac{\alpha \cdot D_{Nozzle}}{\lambda} = G \cdot Re^{\frac{2}{3}} \cdot Pr^{0.42}, \tag{9.20}$$

with

$$G = \frac{d^*(1-2.2d^*)}{1+0.2(h^*-6)d^*} \cdot \left[1 + \frac{10h^* \cdot d^*}{6}\right]^{-0,05}, \tag{9.21}$$

within the following range of validity:

$$0.004 \leq \left(d^{*2} = f\right) \leq 0.04, \tag{9.22}$$

$$2 \leq \left(h^* = \frac{H_{NSD}}{D_{Nozzle}}\right) \leq 12, \tag{9.23}$$

$$2000 \leq Re \leq 100000. \tag{9.24}$$

For a quadratic nozzle arrangement,

$$f = \frac{\pi}{4} \cdot D_{NSD}^2 / L_T^2, \text{ and} \tag{9.25}$$

$$d^* = \sqrt{f} = 0.8862 \frac{D_{NSD}}{L_T}. \tag{9.26}$$

The heat transfer coefficients calculated from Equation (9.20)-(9.26) are compared to values measured with the sensor (Subsection 9.4.1) in Figure 9.17. As a measure of homogeneity the error bars indicate the standard deviation of measurements at 5 different positions on the substrate.

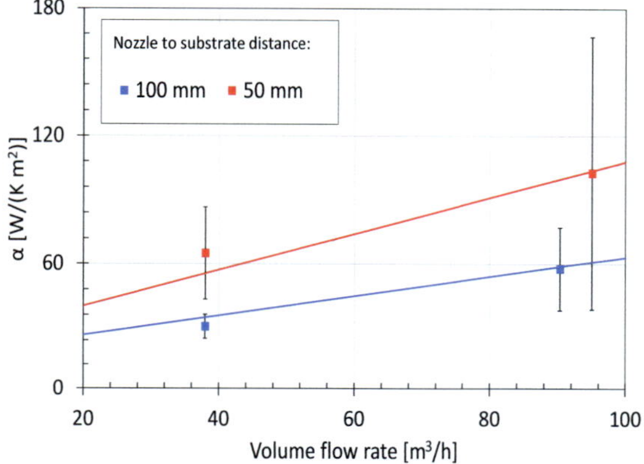

Figure 9.17: *Average heat transfer coefficient as a function of volume flow rate. The squares are values measured and lines are correlations for ARN systems taken from VDI Wärmeatlas (Equation (9.20)-(9.26)).*

As expected, heat transfer coefficients increase with volume flow rate and decrease with nozzle to substrate distance whereas homogeneity decreases with increasing value of the heat transfer coefficient.

Local heat transfer coefficient

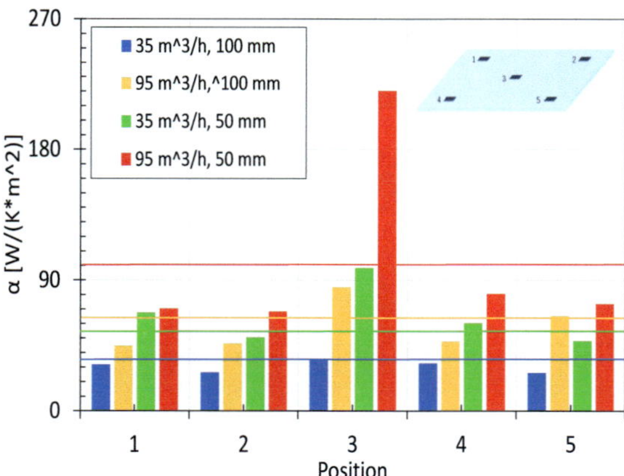

Figure 9.18: *Local heat transfer coefficients as a function of positions within the ARN-drier. Position 3 is in the center of the nozzle field and po-*

sition 1, 2, 4 and 5 are distributed in a radius of 60 mm from the center, towards the corners of the nozzle field (see sketch in top right corner). Average values are indicated by horizontal lines.

Local values of heat transfer coefficient at the five measured positions are illustrated in the bar chart in Figure 9.18.

Especially the value at the center (Position 3) increases at high volume flow rates, indicating that the observed effect is due to large scale inhomogeneity (see Subsection 2.2.3).

Another measure of the homogeneity of local heat transfer was done with the set-up described in Subsection 9.4.2. The change of surface temperature of a plate heated with the ARN-drier is shown for two nozzle to substrate distances in Figure 9.19. At H_{NSD} =20 mm, small scale inhomogeneitys are clearly visible throughout the investigated area. At H_{NSD} =100 mm, large scale inhomogeneities become apparent but an area with homogeneous temperature forms in the center of the substrat.

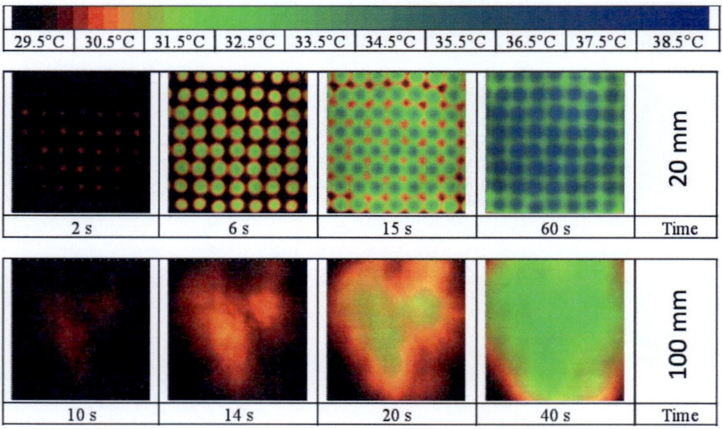

Figure 9.19: Evolution of the temperature distribution on the surface of a plate heated with the ARN-drier for a nozzle to substrate distance of 20 mm and 100 mm.

9.5.2. Array of slot-nozzles drier (ASN-drier)

An experimental characterization of the ASN-drier is currently conducted by Jaiser and Baunach. According to the technical documentation the blowers will provide a maximum volume flow rate of 330 m³/h. The heat transfer coefficients that can be reached for different flow rates are shown in Figure 9.20.

The minimum air temperature is limited as air is heated up when compressed by the drier. Approximate minimum values can be taken from Figure 9.21.

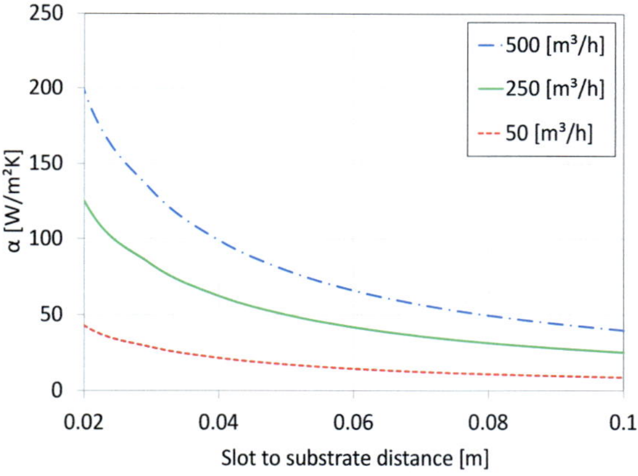

Figure 9.20: Expected heat transfer coefficients of the ASN-drier as a function of slot to substrate distance for three different volume flow rates. The data was computed according to the correlations for ASN in VDI Wärmeatlas chapter Gk.

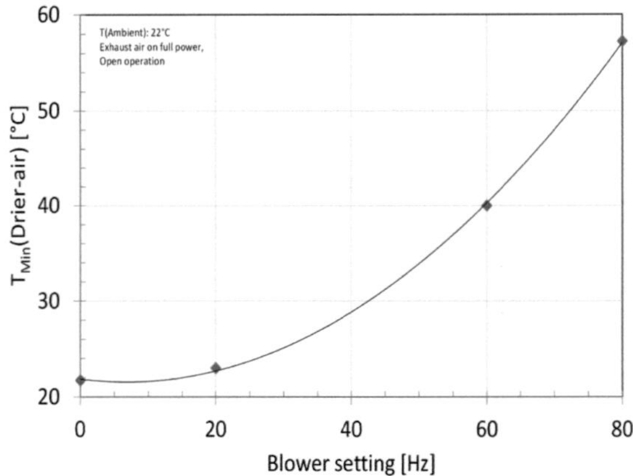

Figure 9.21: Minimum air temperature in the ASN-drier as a function of blower setting. Experimental data was fitted with as 2nd order polynomial function to approximate values in between the measured points.

Due to heat losses between the heating device and drier, the temperature of the support plate and drying air are lower than the set-point value. Calibration measurements and a linear fit are illustrated in Figure 9.22. Note, that the calibration curves can be used for isothermal conditions or as first approximation, only. The exact values are also affected by interaction of support plate, drier air, and ambient air. A permanent measurement of drying air and support plate temperature was implemented and visualized in a LabVIEW routine.

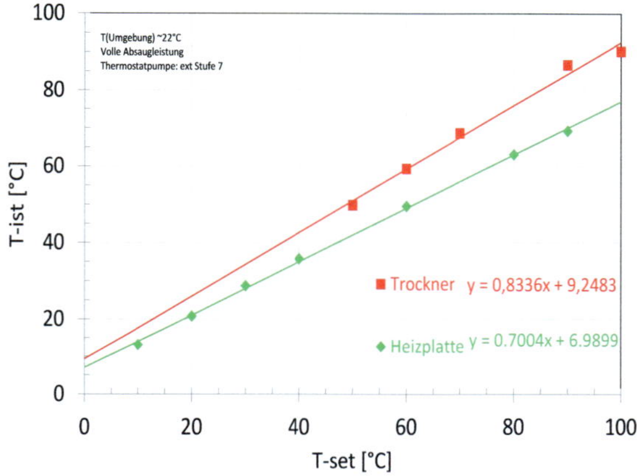

*Figure 9.22: Calibration lines for the temperature of the temperature con-
trolled support plate and the drying air temperature as a function of set-
point temperature.*

9.5.3. Pilot-line slot-nozzle drier (PSN-drier)

Figure 9.23 shows three correlations of heat transfer coefficient and volume
flow rate at room temperature. The volume flow was measured using a
heat-wire-anemometer (TA5, Airflow) and the heat transfer coefficients
were calculated according to an empirical correlations for single slot nozzles (SSN) or arrays of slot nozzles (ASN) by Martin (Martin 2002). The
broken line represents the average heat transfer coefficient according to the
ASN-correlation, whereas the solid line shows local HTC at the sensor positions according to the correlation for SSN.

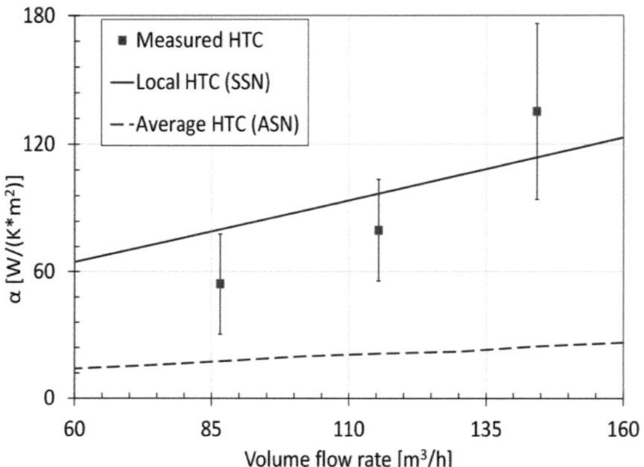

Figure 9.23: Heat transfer coefficient in the pilot line drier as a function of volume flow rate. Computed average and local heat transfer coefficient, are plotted as broken and solid line, respectively. Measurements are indicated by squares.

The measured HTC values (squares) are significantly larger compared to the values computed for an array of slot nozzles. This difference may be due to an insufficient number of measured values (5 measurement positions) or due to the large nozzle spacing. The interaction of adjacent nozzles may be less pronounced so that each nozzle can be described as a single slot nozzle (solid line). This unexpected result highlights the importance of actual measurements of HTCs in technical drying nozzle fields, motivating further research.

The nozzles within each segment are positioned equidistant at 5, 35, 65, and 95 cm from the inlet. Therefore the last nozzle of the first segment and the first nozzle of the second segment are only 10 cm apart, resulting in higher temperature and heat transfer rates at the center of the drier. To compensate for this, the forth nozzle above the substrate (top) was closed. Table 9.5 shows the average nozzle exit velocity of Segment 1 and Segment 2 in this configuration. The average velocity of top and bottom nozzle (1st column) is similar in both segments even though Segment 1 has only 7 open nozzles. The top nozzle velocity in both segments is higher than the bottom nozzle velocity.

Table 9.5:Nozzle exit air velocity for Segment 1and Segment 2 of the pilot line drier. The nozzles above and below the substrate are listed independently in the columns labled top and bottom, repsectively.

	Nozzle exit velocity [m/s]		
Segment	t & b (average)	top	bottom
1 & 2 (average)	2.64	3.18	2.09
1	2.66	3.49	1.82
2	2.62	2.95	2.29

In this configuration, the temperature profile within the drier became more homogeneous. The temperature drop at the front and end is due to cold air entrainment from the environment (Figure 9.24).

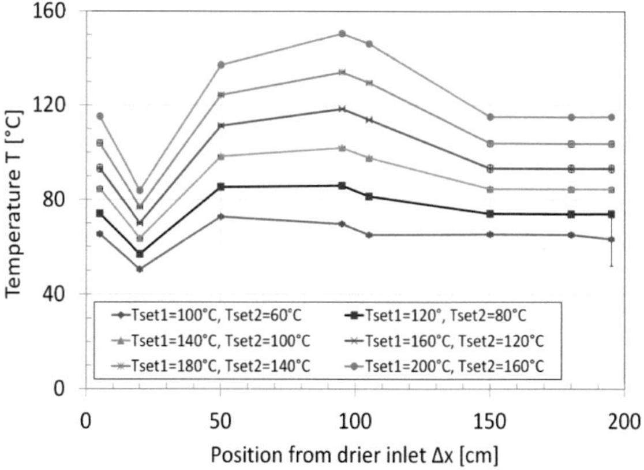

Figure 9.24: Temperature profile within the drier after the forth nozzle of the first segment (at 95 cm) was closed. The set-point temperature of the first segment was set 20°C higher than that in the second. The inlet and outlet blower are operated at 100% and 60% power, respectively. Data

points are connected by piecewise linear fits to emphasize the general trend.

The average temperature in each segment is plotted in Figure 9.25 as a function of set-point temperature. The "error bars" indicate maximum and minimum temperature, especially maximum temperatures have to be considered when using temperature sensitive materials.

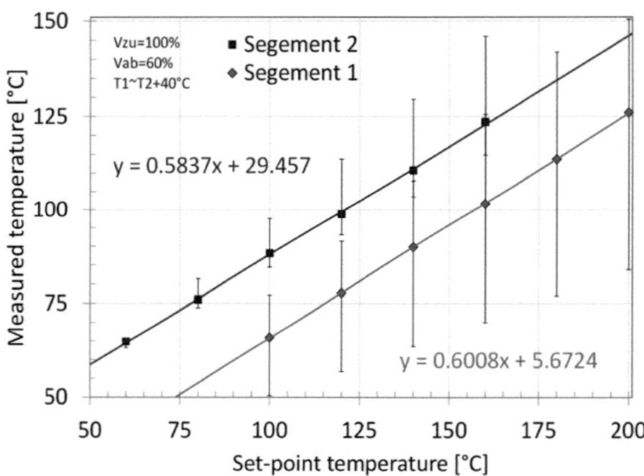

Figure 9.25: Linear calibration curves for the temperature within the two drier segments as a function of set-point temperature.

9.6. Wetting behavior

Though wetting was initially not intended to be a central research topic in this work, wetting behavior turned out to be one of the most important aspects for a successful production of functional films for PSC. The high surface tension and low viscosity of the coating inks result in low Capillary Numbers, and the low wet film thicknesses create a high specific surface area. Under these conditions, wetting behavior plays a major roll for the stability of coated films.

9.6.1. Introduction

The surface tension of a fluid (σ_L) can be determined by an optical analysis of a pendant drop[15]. A fit of the drop shape can be used to calculate its volume V_{Drop} and the surface tension σ follows from the equilibrium of surface and gravitational forces at the "neck", the position where the cross section of the drop has the smallest diameter (D_{Drop}):

$$\sigma_L = \frac{V_{Drop}\rho g}{\pi D_{Drop}},$$ (9.1)

where ρ is the density and g the gravitational acceleration.

The static contact angle of a sessile drop (Θ) on a solid surface was defined by Young as equilibrium of linear forces at the static wetting line:

$$\sigma_S = \sigma_{S,L} + \sigma_L \cos(\Theta),$$ (9.2)

where σ_S is the surface energy of a solid-gas interface and $\sigma_{S,L}$ the surface energy of the solid liquid interface. According the OWRK-Model (Owens and Wendt 1969), the interfacial forces of each phase can be separated into a polar (superscript p) and a disperse (superscript d) part:

$$\sigma_L = \sigma_L^p + \sigma_L^d \quad \text{and} \quad \sigma_S = \sigma_S^p + \sigma_S^d,$$ (9.3)

and the interaction of solid and liquid can be described by:

$$\sigma_{S,L} = \sigma_S + \sigma_L - 2\left(\sqrt{\sigma_L^p \sigma_S^p} + \sqrt{\sigma_L^d \sigma_S^d}\right).$$ (9.4)

Thus the surface energy of a liquid-solid interface is the sum of the surface energies of both phases minus a term proportional to the product of polar and disperse parts in each phase. Eliminating the unknown $\sigma_{S,L}$ in Equation (9.2) with Equation (9.4) gives:

---

[15] A drop hanging from a syringe tip shortly before falling

$$\cos(\Theta) = 1 - \frac{2}{\sigma_L^P + \sigma_L^D}\left(\sqrt{\sigma_L^p \sigma_S^p} + \sqrt{\sigma_L^d \sigma_S^d}\right), \tag{9.5}$$

an equation that describes the static contact angle as a function of disperse and polar part of the surface energy of solid and liquid. For a specific substrate (σ_S^p and σ_S^d are known), lines of constant contact angle can be plotted in diagram showing σ_L^p as a function of σ_L^d. Figure 9.26 shows contact angle of 60°, 45°, and 0° for an ITO-glass substrate without corona treatment. Liquids are characterized by a discrete set of σ_L^p and σ_L^d and are thus represented as points. The $\Theta = 0°$-line is conventionally called "wetting envelope" as all fluid which characteristic points are enveloped by it, are assumed to show "complete wetting".

9.6.2. Surface energy parts of substrates and films

To calculate a wetting envelope, the surface energy and its disperse and polar part have to be known. An overview for typical substrates and films before and after plasma treatment is presented in Table 9.6.

Table 9.6: Over view of surface energy data for typical substates and films. Average values (AVE) and their standard deviation (STDV) are tabulated and if available the disperse (DP) and polar (PP) part. Maximal values achieves by plasma treatment in the batch process or corona treatment in the pilot coating line** presented where relevant.*

Name	SFE (untreated) [mN/m]				SFE (plasma treated / maximal) [mN/m]			
	AVE	STDV	DP	PP	AVE	STDV	DP	PP
Substrates								
PET**	52.07	0.52	45.25	7.69	80.33	0.02	49.44	30.88
Glass *	64.32	0.75	37.12	26.2	77.13	0.78	44.59	32.54
ITO-Glass *	54.55	2.98	41.04	13.5	79.98	0.74	49.28	30.7
Films								
PR-020	54.33	0.43						
P200	29.29	1.45						
PCBM	26.79	1.97						
P200:PCBM **	29.69	0.11	28.61	1.08	42.53	1.23	30.16	12.36
VPAI 40.83	46.48	1.04						
ZnO-alpha	40.43	1.97						

Using the values from Table 9.6 and Equation (9.5), lines of constant contact angle can be computed.

Figure 9.26 shows lines for contact angles of 60°, 45° as well as the 0° wetting envelope for untreated ITO-glass. Data points for different ink compositions can now be plotted as shown exemplarily for pure PEDOT:PSS (VPAI type, open triangle), PEDOT:PSS with an addition of 10 wt.% methanol (filled triangle) and pure methanol (square).

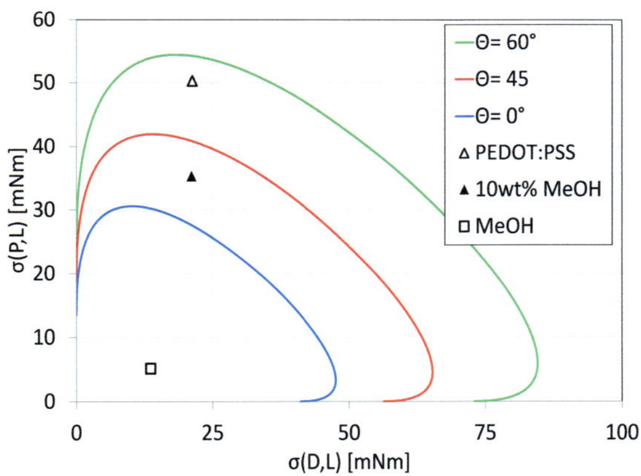

Figure 9.26: Lines of constant contact angle for an untreated ITO-glass substrate. Liquids are characterized by a point in this diagram.

9.6.3. Surface tension parts of solvents and inks

A list of polar and disperse parts of the surface tension of various solvents is presented Table 9.7.

Whereas the surface tension of organic solvents and alcohols typically ensures good wetting behavior, a polar part of 54 MN/m for water can often cause de-wetting or the breakup of films. As discussed before (Section 4.3 and Section 4.4) alcohols or surfactants can be added to water based inks to improve wetting behavior. Surfactants are indispensable for high conducting PH1000/DMSO dispersions but alcohols are sufficient to ensure wetting of hole conducting PEDOT:PSS types without DMSO.

Table 9.7: Overview of disperse and polar part of the surface tension of various solvents. Values were taken from the Studienarbeit of Daniel Griese (Griese 2011) and own measurements.*

Name	σ [mN/m]	σP [mN/m]	σD [mN/m]
Water / Alcohols			
VPAI 40.83	69.2	53.6	15.6
Wasser	72.8	50.8	21.8
Methanol	18.9	5.1	18.9
Ethanol*	22.6	6.4	16.2
Isopropanol*	21.0	2.6	18.4
Organic solvents			
Chlorobenzene*	32.0	5.5	26.5
o-Xylene*	29.3	3.0	26.26
Chloroform*	26.9	2.4	24.5
Toluene*	28.4	2.3	26.1

Figure 9.27 shows the polar and disperse part mixtures of water with methanol (a) and ethylene glycol (b).

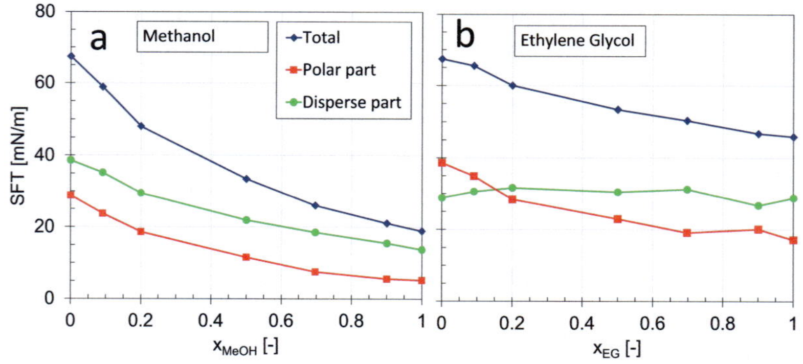

Figure 9.27: Polar and disperse part of the surface tension of water methanol (a) and water ethylene glycol (b) mixture.

Whereas methanol has a much lower disperse part compared to water, the disperse part of ethylene glycol – Water mixtures remains constant for all mixing ratios. It is thus possible to achieve a desired position relative to the wetting envelope by mixing of different solvents. In contrast, a plasma treatment will mostly affect the polar part of a substrates surface energy (Wenz 2012).

9.6.4. Evolution of the surface energy of activated substrates

The increase of surface energy or activation of a substrate or solid film was done by plasma treatment in an oxygen plasma generator for the batch coater process, or in an inline corona generator for the pilot coating line. Whereas the time between activation and coating is less than 5 min in the PC even at low coating speeds, longer times may occur during manual handling of the substrates in the BC process. The evolution of surface energy as a function of time after activation is plotted in Figure 9.28. Surface energy decreases linear over time and reaches half of the increase after approximately 90 min.

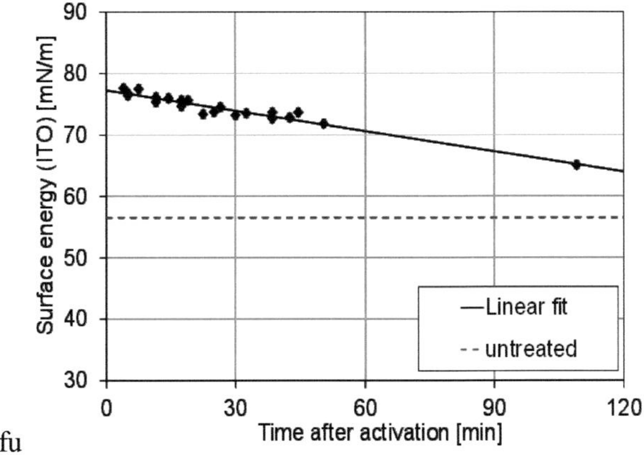

Figure 9.28: Surface energy of oxygen plasma treated ITO-glass substrates as a function of time after the activation. The slope of the linear decrease depends on ambient conditions e.g. relative humidity.

The rate at which the surface energy decreases depends on ambient conditions and was not investigated in detail here. However it is important to note that the "deactivation" process can occur over time spans of minutes to hours and thus has to be taken into account during experimental work. In this work plasma treatment was done less than 5 min before coating and surface energy was assumed to be at its maximum value.

9.7. Material properties (drying)

Table 9.8:Material properties for the calculation of film temperature and drying process, taken from Yaws (Yaws 1999).

Name	Chemical structure	Molar mass	Boiling point	Density	Enthalpie of evaporation	Diffusional volume	Antoine parameters		
		g/mol	@ 1 bar °C	@ 20°C g/cm	kJ/kg		A	B	C
Chlorbenzol	C6H5Cl	112.6	132	1.11	339.24	109.65	7.17	1549.20	229.26
m-Xylol	C8H10	106.2	144	0.87	392.71	132.00	4.14	1463.22	-57.99
o-Dichlorbenzol	C6H4Cl2	147.0	179	1.30	312.47	128.34	6.92	1538.30	200.00
Indan	C9H10	118.2	178	0.97	386.37	147.90	6.11	1577.32	-66.83
Chloroform	CHCl3	119.4	61	1.48	265.40	81.21	4.57	1486.46	-8.61
Toluol	C7H8	92.1	111	0.87	400.13	111.48	6.95	1344.80	219.48
Water	H2O	18.0	100	1.00	2260.00	13.10	8.14	1810.94	244.49
Isopropanol	C_3H_8O	60.1	82	0.79	794.20	70.82	4.58	1221.42	-87.47
Dimethylsulfoxide	C_2H_6OS	78.1	189	1.10	677.08	67.36	4.49	1807.00	-61.00

Table 9.9: Temperature dependent material properties of air at 1 bar pressure taken from VDI Wärmeatlas (VDI 2002).

T	c_p	λ	ν	Pr	ρ	η
°C	kJ/(kg K)	mW/(m K)	m²/s	-	kg/m^3	Pa*s*10^(-6)
0	1.0059	24.36	0.0000135	0.7179	1.275	17.218
10	1.0061	25.12	0.0000144	0.7163	1.230	17.715
20	1.0064	25.87	0.0000153	0.7148	1.188	18.205
30	1.0067	26.62	0.0000163	0.7134	1.149	18.689
40	1.0071	27.35	0.0000172	0.7122	1.112	19.165
50	1.0077	28.08	0.0000182			19.635
60	1.0082	28.80	0.0000192	0.7100	1.045	20.099
70	1.0089	29.52	0.0000203			20.557
80	1.0097	30.01	0.0000213	0.7083	0.986	21.009
90	1.0105	30.93	0.0000224			21.455
100	1.0115	31.39	0.0000235	0.7070	0.933	21.896
120	1.0136	32.75	0.0000257	0.7060	0.885	22.763
140	1.0160	34.08	0.0000280	0.7054	0.843	23.610
160	1.0188	35.39	0.0000304	0.7050	0.804	24.439
180	1.0218	36.68	0.0000329	0.7049	0.768	25.251

9.8. Technical annealing of functional films

Similar to the effects observed in the drying process, a subsequent annealing may change the properties of functional films. In general, the mobility of molecules and particles within the film is increased during annealing, enabling desorption of residual solvents or a structural rearrangement of polymers, molecules, or particles. The relevant process parameters are temperature, treatment-duration, and atmospheric conditions. Canonically annealing processes are either classified as thermal annealing - under inert or ambient atmosphere or solvent assisted annealing –in an atmosphere that contains one or more solvents. Annealing has been reported to improve the performance of polymer-fullerene photoactive layers (Li, et al. 2005; Miller, et al. 2008; Pearson, et al. 2012), hybrid photoactive layers (Wu and Zhang 2010; Dixit, et al. 2012), PH1000 electrodes (Huang, et al. 2005), and silver electrodes (Walker and Lewis 2012). Solvent annealing is often done by covering the film with a petri dish (e.g. (Miller, et al. 2008; Wu and Zhang 2010)). Though this method is very simple, solvent concentration is unknown and the obtained results are thus difficult to scale to a technical process.

In Subsection 9.8.1 the impact of solvent content on the properties of functional films is shown, using water in PH1000 films as an example. The impact of thermal and solvent-assisted annealing on organic and hybrid photoactive layers and electrodes is addressed briefly in the following subsections for exemplary material systems. A more detailed investigation of individual process parameter is documented in the Bachelor thesis of Jana Kumberg (Kumberg 2012).

9.8.1. Water absorption into PH1000 electrode films

The slow rate at which solvents desorb from functional films was first observed when measuring the specific conductivity of a PH1000 film that was taken from ambient conditions and introduced into inert conditions (see Figure 6.5). In the same set-up, dry nitrogen was mixed with saturated nitrogen. The resulting relative humidity in the gas phase and the specific conductivity of the film were measured after ~10 minutes residence time at different mixing ratios (Figure 9.29).

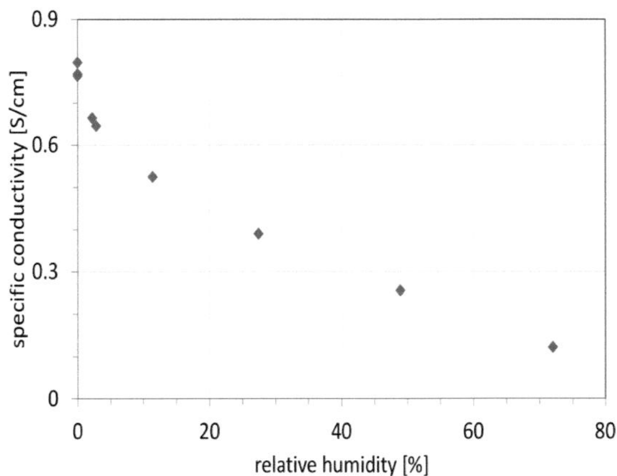

Figure 9.29: Specific conductivity of a 200 nm PH1000 film (processed without DMSO) as a function of humidity. As the film swells, conductivity decreases.

Though the residence time is not sufficient to ensure equilibrium conditions, the strong impact of residual water on the properties of a functional film becomes obvious. On the one hand, the dependence of conductivity on relative humidity may be utilized in other applications for example as a cheap, printable humidity sensor. But on the other hand, residual water, or residual solvents have to be removed from functional films in PSCs by an annealing process before encapsulation.

9.8.2. Temperature limitation

Up to the limit of thermal degradation, a higher annealing temperature typically has a positive effect on the investigated figure of merit. As an example, the specific conductivity of a silver electrode layer on PET is shown in Figure 9.30. In the temperature range from 100 to 180 °C, the conductivity increases strictly monotonic. As a reference, the conductivity of bulk silver is 614 kS/cm indicating the large potential of annealing temperature.

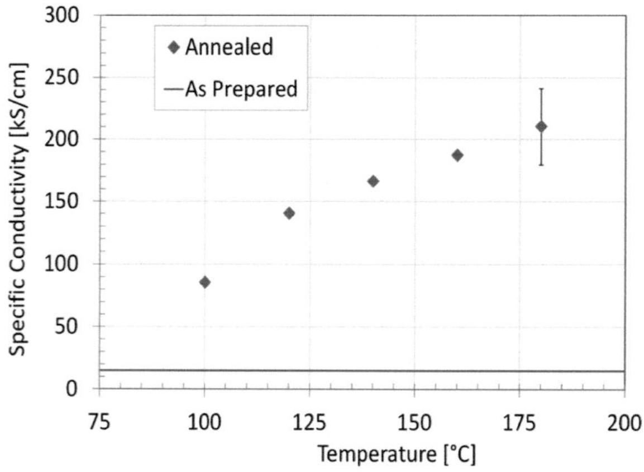

Figure 9.30: Specific conductivity of ~200 nm silver nanoparticle films after thermal annealing (30 min) at different temperatures.

The standard deviation, plotted as error bars, increases at high temperatures due to a thermal degradation of the PET substrate. Figure 9.31 shows a photo of samples after annealing at different temperatures. At 140°C and above, the PET substrate is deformed due to thermal shrinkage.

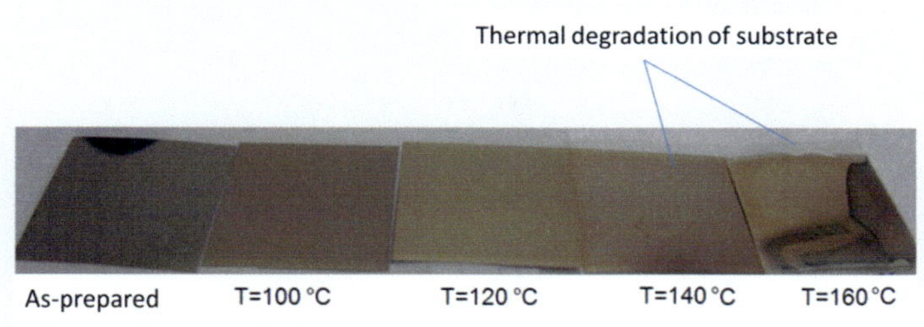

Figure 9.31: Photo of Ag-Pr020 layers on PET substrate after annealing.

A similar increase in specific conductivity and standard deviation was observed for PH1000 films (Kumberg 2012). In photoactive films, the organic compounds themselves degrade above ~140°C.

The annealing temperature of a continuous process for PSCs is thus limited to less than 140°C due to thermal instability of photoactive materials and substrates.

9.8.3. Time dependence of annealing processes

At lower temperatures, a lower residual solvent content can be achieved by longer annealing times. Figure 9.30 shows the specific conductivity of PR020 films as a function of residence time on a hot plate at 120°. Though the highest increase occurs before the first measured value at 30 min, a continuous increase in conductivity can be observed even after 90 min. Similar time spans of 30 – 60 minutes were observed for the effect of annealing time on the conductivity of PH1000 electrode films and the light absorption of P3HT:PCBM and P3HT:QD photoactive films (see (Kumberg 2012)).

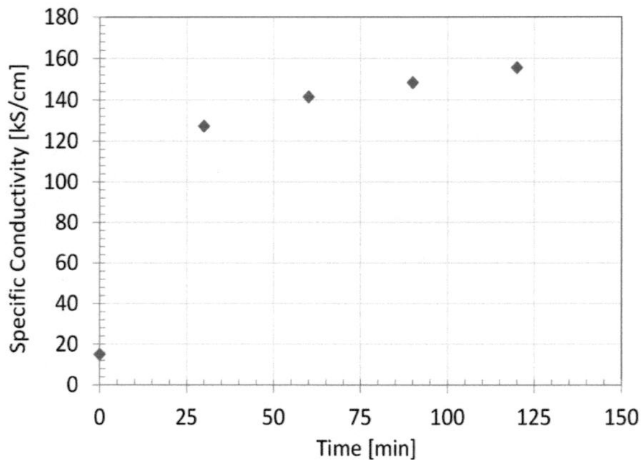

Figure 9.32: Specific conductivity of a silver nanoparticle layer (PR020) as a function of annealing time on a hot plate at 120°C in nitrogen atmosphere.

Whereas the initial increase is probably due to sintering of the nanoparticles, the continuous slow increase can be caused either by desorption of residual solvents or by structural changes of the film morphology.

Annealing times in the order of minutes to hours are difficult to realize in a continuous technical process. A practical solution may be to anneal coated rolls one by one in an oven before transfer to the next continuous coating step. Another method to increases the rate of mass transfer is the treatment with an atmosphere that is saturated with a solvent with a higher desorption rate. This so called "solvent annealing" could also increase the rate at which structural changes occur.

9.8.4. Solvent annealing

P3HT-P200:PCBM layers were solvent annealed in a sorption chamber that was flushed with a Nitrogen Chlorobenzene mixture at an activity of 0.72 and a cell temperature of 25°C. The average specific light absorption of the samples is plotted in Figure 9.33 as a function of residence time within the sorption chamber. The absorption increases by 10% within the first 30 minutes and reaches a maximal increase at 120 min. No clear increasing tendency can be observed after 30 min.

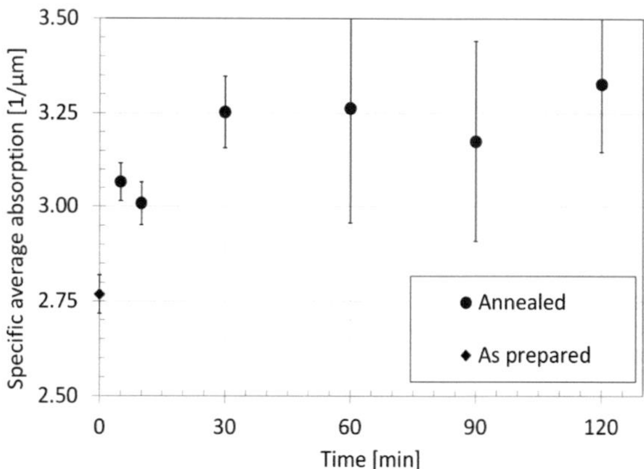

Figure 9.33: Specific average absorption from 450-650 nm as a function of residence time in the sorption cell at 25°C and a solvent (Chlorobenzene) activity of 0.72. All samples were annealed for 10 min at 110°C in nitrogen atmosphere prior to characterization.

Though the experiment shows that solvent annealing can be applied to change the functional properties of semi-conducting films, the increase still required long residence times (~30 min). However, solvent annealing provides an additional degree of freedom for the process optimization. Annealing with solvents that have a higher or a selective solubility i.e., could be used to optimize morphologies of films prepared with non-halogenated solvent inks.

9.8.5. Annealing of hybrid photoactive layers

Annealing is even more important for hybrid photoactive layers. Nanoparticles have to be stabilized by ligands in order to inhibit agglomeration in the coating ink. During drying, some ligands can be removed from the ink but the annealing step determines the residual ligand content in the dry film (see Figure 9.34).

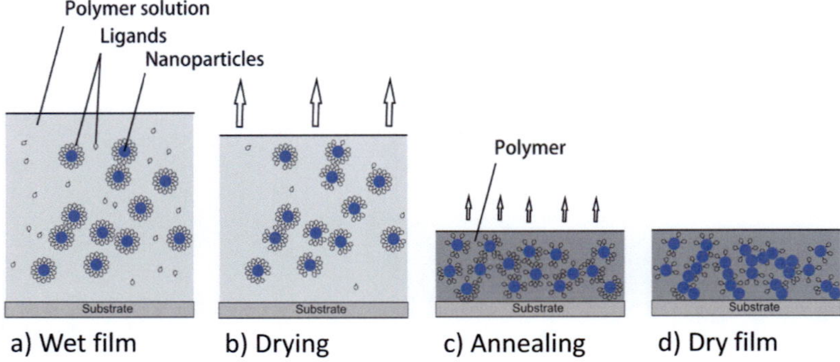

Figure 9.34: Schematic representation of the ligand desorption during drying and annealing of hybrid photoactive films.

The strong impact of annealing conditions is shown by the increase of light absorption for hybrid films, treated with different annealing temperatures and residence times (see Figure 9.35).

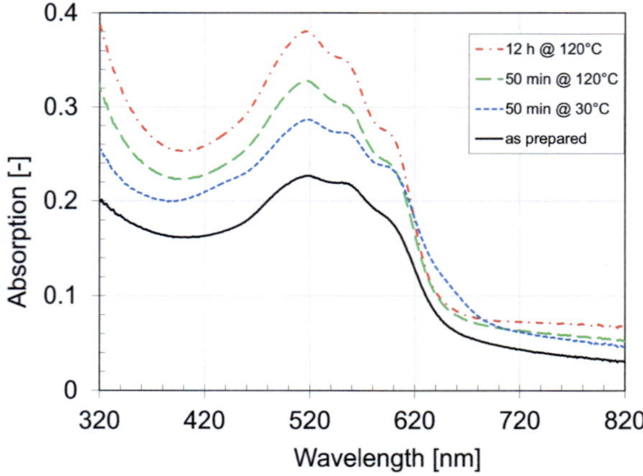

Figure 9.35: Absorption spectra of hybrid photoactive films before annealing (solid line) and after thermal annealing in a vacuum drying cabinet at 10 mbar for variable residence time and temperature.

Both, higher temperature and longer residence times facilitate the desorption of ligands significantly and result in better optical properties.

In summary: residual solvents, water, or ligands can influence the properties of functional films significantly and have to be removed by an annealing process step. Annealing temperature is limited to values below 140°C due to the stability of substrates and photoactive materials. At typical temperature of 120°C, desorption processes in functional films with nanometer thickness occur in time spans of minutes to hours. Therefore long residence times are required for the annealing process. Though initially only moderate improvements were achieved, solvent annealing may provide methods to speed up the solvent desorption and morphological changes.

9.9. Recommended process parameters

The parameters proposed in this section are based on the results and experiences gained in this thesis. As discussed in the outlook section there is room for further improvement and this parameter set is meant as starting point for further optimization.

The preparation of inks for standard and inverted architecture cells is summarized in Table 10.10. A surfactant content of 0.5 wt.-% is recommended for the PH1000 dispersion. For the P3HT:PCBM ink, a polymer concentration of 12 mg/ml and a solvent composition with 80 v.-% low boiling point solvent is recommended when processing at room temperature. Ink composition and preparation is tabulated in Table10.10.

Table 9.10: Ink preparation for coating inks in standard architechture (P3HT-P200:PCBM, VPAI40.83) and inverted achitecture (Ag-PR020, ZnO-gen, P3HT-P200:PCBM, PH1000).

System	Ink preparation
Ag-PR020	Diluted 1:1 by volume with IPA; 10 min ultrasound treatment
ZnO-gen	Diluted with IPA 1:1 by volume
P3HT-P200:PCBM	12 mg/ml P3HT, 9.6 mg/ml PCBM in 80v.-%Tol, 20v.-% Ind or 80 v.-%CF, 20 v.-% DCB; stirred on 60°C hot plate for 24 h
PH1000	90 v.-% PH1000+ 5 v.-% DMSO+ 5 v.-%Water-Surfactant (0.5 wt.-% TX100); 10 min ultrasound treatment
VPAI40.83	44 v.-% VPAI40.83 + 44 v.-%Water+12v.%MeOH ; 10 min ultrasound treatment

The process parameters given in Table 9.11 are based on room temperature (20°C-25°C) coating of typical dry film thicknesses for the individual layers. The density of the film is calculated assuming an ideal mixture of solid and liquid. Note that densities are taken from literature and may differ from actual material density within the film. To utilize the effect of shear induced crystallization within the photoactive layer, while remaining sufficiently far from the stability limit, a coating velocity of 5 m/min and a relative low gap of 38μm were chosen.

Table 9.11: Overview of independent and calculated process parameters for typical layers within a polymer solar cell. Material properties are calculated at room temperature (20°C).

System			Ag-PR020	ZnO-gen	P3HT:PCBM	PH1000	VPAI40.83
Independent parameters	Symbol	Unit					
Coating velocity	v	m/min	5	5	5	5	5
Dry film thickness	d	nm	200	40	100	200	40
Density solid	ρS	g/cm³	10.5	5.6	1.3	1.6	1.6
Solid content	x	wt.-%	17.50	0.72	1.93	1.77	0.60
Density of solvent	ρLM	g/cm³	0.78	0.78	1.10	1.00	1.00
Solvent Viscosity	$\eta\infty$	mPas	1.96	1.96	0.77	1.00	1.00
Gap width	G	µm	38	38	38	38	38
Surface tension	σ	mNm	33	33	30	30	30
Coating width	B	mm	35	35	35	35	35
Dependent variables	Symbol	Unit					
Density of (wet) film	ρLM	g/cm³	0.93	0.78	1.10	1.01	1.00
Wet film thickness	D	µm	12.88	39.83	5.89	17.93	10.24
Volume flow rate	V	ml/min	2.25	6.97	1.03	3.14	1.79
Dimensionless gap width	G^*	µm	2.95	0.95	6.45	2.12	3.71
Shear rate (in the gap)	γ	1/s	2193	2193	2193	2193	2193
Viscosity	η	mPas	9.61	2.01	1.85	9.46	2.32
Capillary Number	Ca	-	0.0243	0.0051	0.0052	0.0264	0.0065

Capillary Numbers and dimensionless gap widths for the parameters defined in Table 9.11 are plotted in Figure 9.36.

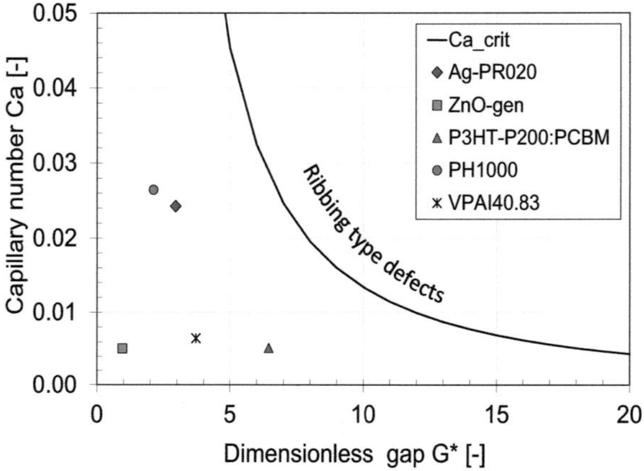

Figure 9.36: Position of the recommended operating points for different layers in PSCs within the Ca-G diagram. All points are well within the stable region.*

A lower coating speed and higher gap width may be advisable under certain conditions. In a batch coating process, the handling of pump and slot-die is easier at coating speeds below 2.5 m/min. When coating over taped in ITO-sheets or rigid glass substrates, a higher slot width of ~50 µm provides more tolerance to positioning errors.

9.10. Complementing experimental data

9.10.1.   Measurement of coating speed

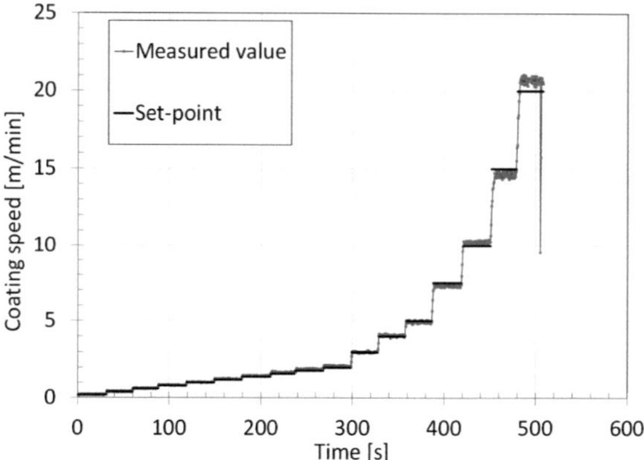

Figure 9.37: Fluctuation of line speed for velocity steps over the complete operating range from 0.2 m/min to 20 m/min. The re-winder was set as speed master. The sample rate was set to 500 ms resulting in a definition of 0.0064 m/min.

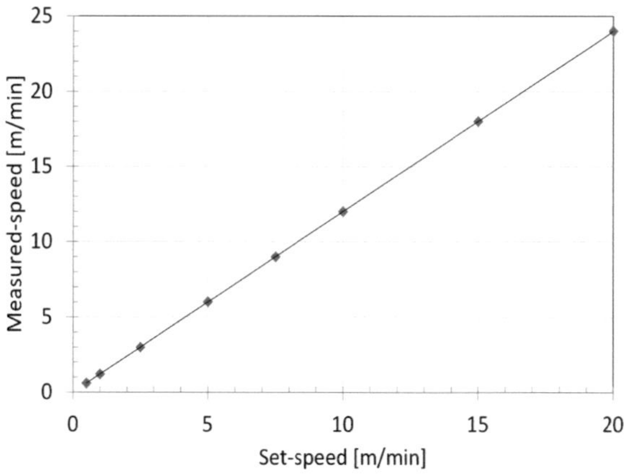

Figure 9.38: Absolute speed measured with an SMT-500C (Aluris GmbH & Co. KG) manual digital tachometer is 20% higher than the set-point in when using the gravure roll as speed master.

9.10.2. Baselines for absorption measurements

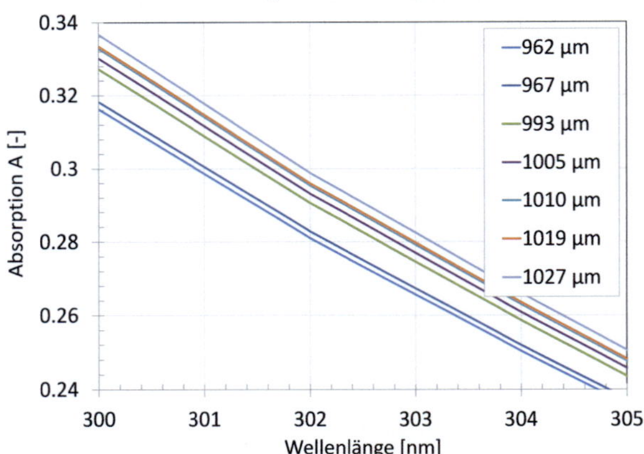

Figure 9.39: Light absorption spectra of glass substrate in the UV regime for different substrate thicknesses. Integral values of these curves are shown in the next figure.

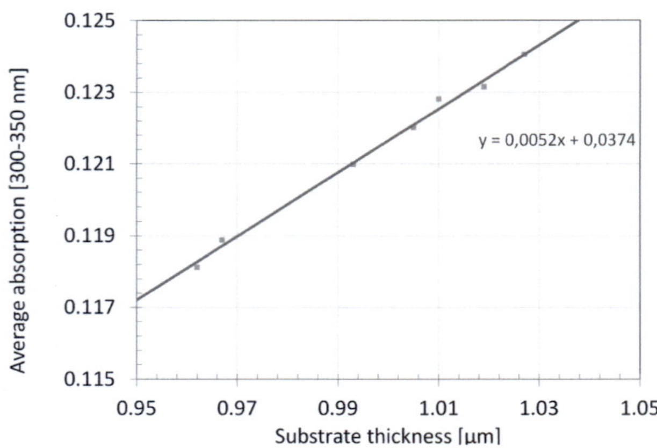

Figure 9.40: Correlation of average light absorption and substrate thickness.

9.10.3. Surface tension of PCBM solutions

In Section 4.3 the surface tension of coating inks was described as a function of solvent composition only. Different solvent compositions will lead to changes in surface energy of several milli-newton per meter. The impact of solid content is less pronounce as shown by the decrease in surface tension as a function of PCBM concentration in Figure 9.41.

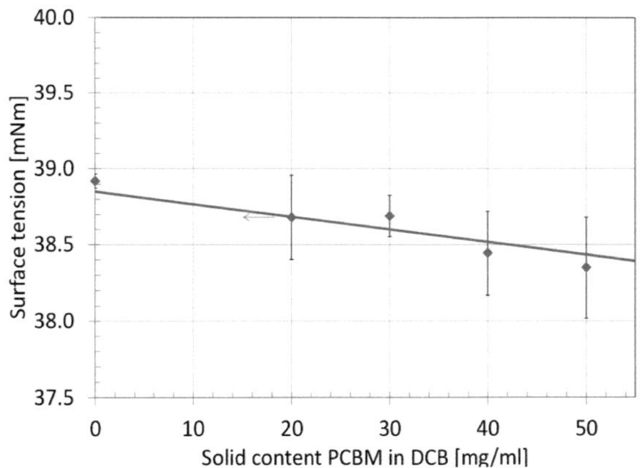

Figure 9.41: Surface tension of PCBM dissolved in Dichlorobenzene as a function of solid content.

While the decrease of less than 1 mN/m over the complete range of solubility is insignificant for the absolute value used e.g. in coating window calculations, the relative difference due inhomogeneous drying may be a driving force for surface tension driven flow and cause film instability during drying.

9.10.4. Layer thickness calibration for knife coating

As discussed in Subsection 5.3.2, film thickness for knife coating has to be calibrated for each substrate fluid material system as a function of coating velocity. Calibration curves for typical inks on glass substrates are presented in Figure 9.42. The substrates for PH1000 films were plasma treated (for 2 min at 80% power). All other substrates were coated after the standard cleaning procedure. Details on the inks composition can be found in Table 10.12.

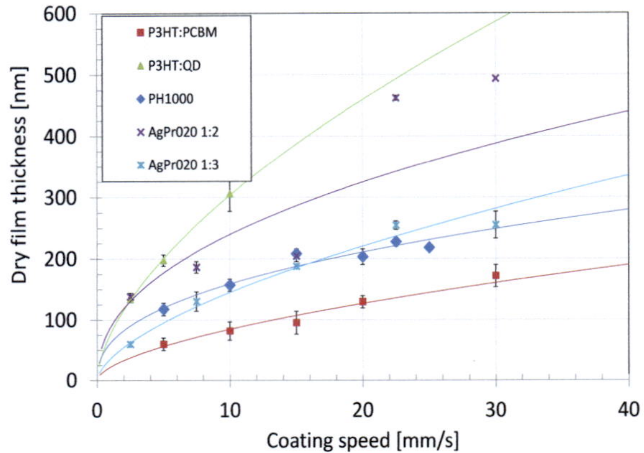

Figure 9.42: Calibration curves for knife coating of functional films. Coating was done with 100 µl volume, at 100 µm knife gap and a temperature of 40 °C.

The dry film thickness (d) can be fitted empirically as a function of coating speed (u):

$$d = a_{knife} u^{n_{knife}}.$$

(9.6)

The wet film thickness can be approximated as

$$D = d \cdot \frac{\rho_{Film}}{\rho_{solid}} \frac{1}{x_{solid}} = d \cdot c.$$

(9.7)

The fit parameters a_{knife} and n_{knife} and the ink specific constant c are tabulated in Table 10.12.

Table 9.12: Summary of composition and fit parameters for the film thickness calibration curves presented in Figure 9.42.

Name	Composition	c=D/d [-]	a_{knife} [nm]	n_{knife} [-]
P3HT-P200:PCBM	6mg/ml P3HT: 4.8 mg/ml PCBM in CB	117.86	22.04	0.5839
P3HT-M:QD	6mg/ml P3HT: 4.8 mg/ml PCBM in 95 v.-%CB, 5v.% PYR	16.16	76.58	0.5982
PH1000	8 part PH1000 diluted with 1 part MeOH (by volume)	172.34	62.23	0.4076
AgPr020-1:2	1 part PR020 diluted with 2 parts IPA (by volume)	0.80	87.77	0.4369
AgPr020-1:3	1 part PR020 diluted with 3 parts IPA (by volume)	1.04	36.11	0.6043

9.10.5. Gravure coating of the organic dummy system

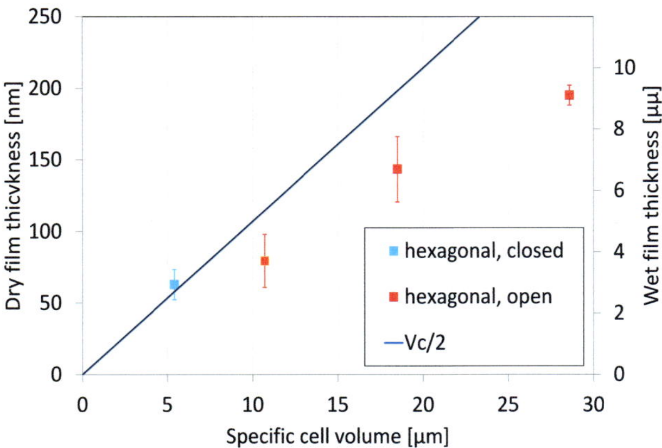

Figure 9.43: Dry and wet film thickness of a 3 wt.% solution of the organic dummy system (PS in XYL) for different specific cell volumes in the stiff blade gravure coating regime.

9.10.6. Film surface topography for different coating methods

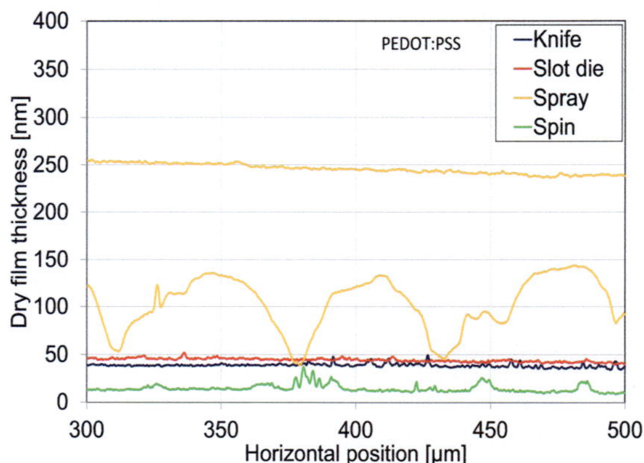

Figure 9.44: Surface topography of VPAI 40.83 films prepared by knife, slot-die, spray, and spin coating. Two layer thicknesses are shown for knife coating: wheras for thin films the individual droplets of the spray can be seen in the dry film, they coalesce at higher (e.g. 250nm) film thickness. Films with good homogeneity at the desired film thickness of less than 50 nm were produced by knife, slot-die, and spin coating.

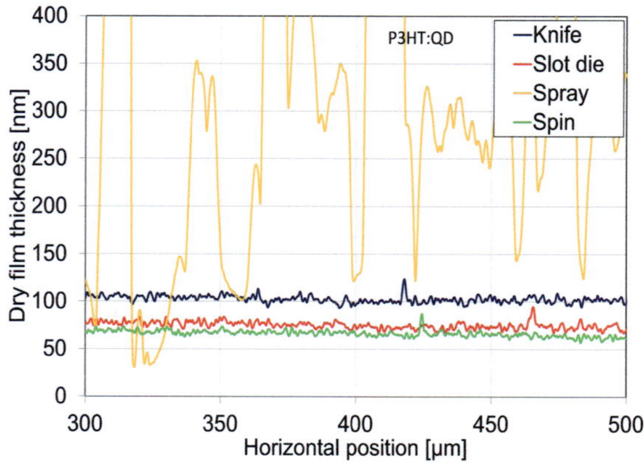

Figure 9.45: Surface topography of P3HT-M:QD films prepared by knife, slot-die, spray, and spin coating. Films with good homogeneity at the

desired film thickness of ~100 nm were produced by knife, slot-die, and spin coating.

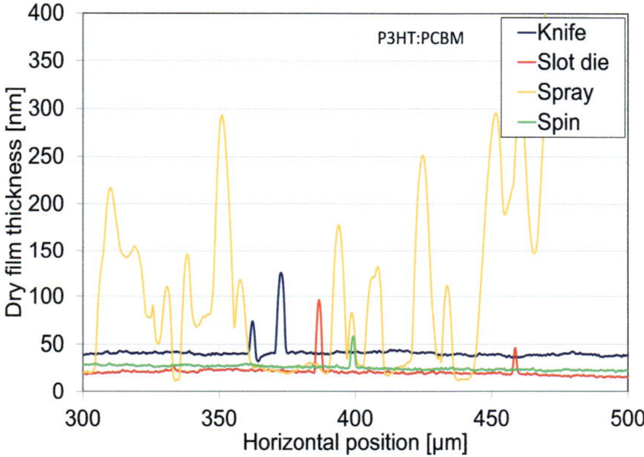

Figure 9.46: Surface topography of P3H T-M:PCBM films prepared by knife, slot-die, spray, and spin coating. Films with good homogeneity at the desired film thickness of ~50 nm were produced by knife, slot-die, and spin coating.

9.10.7. Drying of hybrid photoactive layers

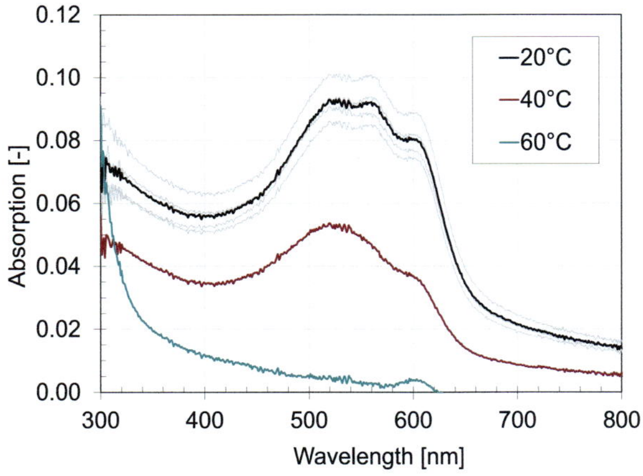

Figure 9.47: Absorption spectra of films that were dried with an identical drying air speed of 0.3 m/s and different isothermal temperatures. Spectra

of individual samples are shown in grey for the 20°C-sample as an indication for the statistical deviation. Drying at low temperatures significantly increases light absorptions of the prepared samples.

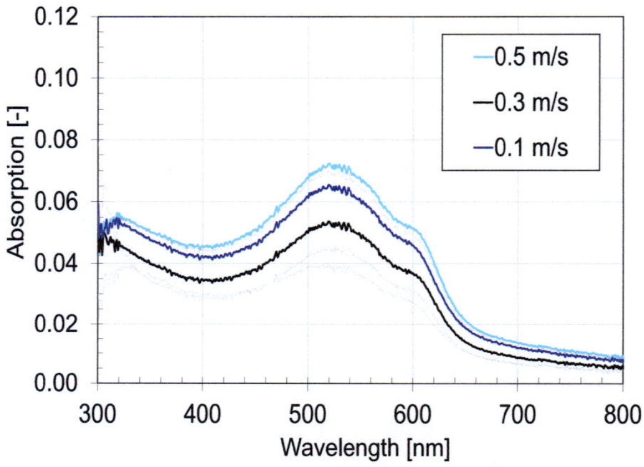

Figure 9.48: Absorption spectra of films dried at different drying air speeds and constant temperature. Spectra of individual samples are shown in grey for the 0.3m/s-sample as an indication for the statistical deviation. Air flow rate has no significant impact on light absorption.

10. References

Aernouts, T., T. Aleksandrov, et al. (2008). "Polymer based organic solar cells using ink-jet printed active layers." Applied Physics Letters **92**: 033306.

Angmo, D., M. Hösel, et al. (2012). "All solution processing of ITO-free organic solar cell modules directly on barrier foil." Solar Energy Materials and Solar Cells.

Blankenburg, L., K. Schultheis, et al. (2009). "Reel-to-reel wet coating as an efficient up-scaling technique for the production of bulk-heterojunction polymer solar cells." Solar Energy Materials and Solar Cells **93**(4): 476-483.

Brabec, C. J. (2004). "Organic photovoltaics: technology and market." Solar Energy Materials and Solar Cells **83**(2): 273-292.

Brabec, C. J. (2008). Organic photovoltaics materials, device physics, and manufacturing technologies. Weinheim, Wiley-VCH-Verl.

Brabec, C. J., A. Cravino, et al. (2001). "Origin of the open circuit voltage of plastic solar cells." Advanced Functional Materials **11**(5): 374-380.

Buss, F., B. Schmidt-Hansberg, et al. (2013). Trocknung und Strukturbildung von Polymersolarzellen (Poster). Jahrestreffen der Fachgruppe Trocknungstechnik Magdeburg, Germany.

Cavadini, P., J. Krenn, et al. (2012a). "Investigation of surface deformation during drying of thin polymer films due to Marangoni convection." Chemical Engineering and Processing: Process Intensification.

Cavadini, P., W. Schabel, et al. (2013a). Device for transferring heat or mass, comprising hexagonal jet nozzles, and method for treating surface layers., WO Patent 2,013,045,026.

Cavadini, P., P. Scharfer, et al. (2012b). Investigation of heat transfer within an array of impinging jets with local extraction of spent fluid. 16th International Coating Science and Technology (ISCST) Symposium, Atlanta, GA.

Cavadini, P., P. Scharfer, et al. (2013b). Numerical and experimental investigation of the heat transfer coefficient distribution at target plate under an array of impinging jets with local removal of the spent fluid. 10th European Coating Symposium, Mons, Belgium.

Chang, C.-C., C.-L. Pai, et al. (2005). "Spin coating of conjugated polymers for electronic and optoelectronic applications." Thin Solid Films **479**(1): 254-260.

Chang, Y. H., S. R. Tseng, et al. (2009). "Polymer solar cell by blade coating." Organic Electronics **10**(5): 741-746.

Chen, H.-Y., J. Hou, et al. (2009). "Polymer solar cells with enhanced open-circuit voltage and efficiency." Nat Photon **3**(11): 649-653.

Coatema-GmbH (2010). Base Coater - Manual - Volume 1 and 2. Dormagen.

Colsmann, A. (2008). Ladungstransportschichten für effiziente organische Halbleiterbauelemente. Karlsruhe, Universitätsverlag Karlsruhe.

Colsmann, A., A. Puetz, et al. (2011). "Efficient Semi-Transparent Organic Solar Cells with Good Transparency Color Perception and Rendering Properties." Advanced Energy Materials **1**(4): 599-603.

Connors, K. A. and J. L. Wright (1989). "Dependence of surface tension on composition of binary aqueous-organic solutions." Analytical Chemistry **61**(3): 194-198.

Conti, J. and P. Holtberg (2011). "International energy outlook 2011." US Energy Information Administration **5**.

10. References

Cros, S., R. De Bettignies, et al. (2011). "Definition of encapsulation barrier requirements: A method applied to organic solar cells." Solar Energy Materials and Solar Cells **95**: S65-S69.

Darmstädter, P. (2012). Konstruktion und Inbetriebnahme eines Gravurwalzenmoduls zur Beschichtung halbleitender Filme für polymerbasierte Solarzellen. Institut für Thermische Verfahrenstechnik, Karlsruher Institut für Technologie.

De Gennes, P.-G. (1985). "Wetting: statics and dynamics." Reviews of Modern Physics **57**(3): 827.

Dennler, G., C. Lungenschmied, et al. (2006). "A new encapsulation solution for flexible organic solar cells." Thin Solid Films **511**: 349-353.

Dennler, G., M. C. Scharber, et al. (2009). "Polymer-Fullerene Bulk-Heterojunction Solar Cells." Advanced Materials **21**(13): 1323-1338.

Dimitriev, O. P., D. A. Grinko, et al. (2009). "PEDOT:PSS films—Effect of organic solvent additives and annealing on the film conductivity." Synthetic Metals **159**(21–22): 2237-2239.

Dixit, S. K., S. Madan, et al. (2012). "Bulk heterojunction formation with induced concentration gradient from a bilayer structure of P3HT: CdSe/ZnS quantum dots using inter-diffusion process for developing high efficiency solar cell." Organic Electronics **13**(4): 710-714.

Döll, H. (2009). Development Coater - Technical documentation. T.-T. Engineering. Murgenthal.

Döll, H. (2013). Angabe zu ungefährer Genauigkeit des Lippen-Substratabstand.

Durst, F. and H.-G. Wagner (1997). Slot coating. Liquid Film Coating, Springer: 401-426.

Duvivier, D., T. D. Blake, et al. (2013). "Toward a Predictive Theory of Wetting Dynamics." Langmuir **29**(32): 10132-10140.

Eastman, S. A., S. Kim, et al. (2012). "Effect of Confinement on Structure, Water Solubility, and Water Transport in Nafion Thin Films." Macromolecules **45**(19): 7920-7930.

Eastoe, J. and J. S. Dalton (2000). "Dynamic surface tension and adsorption mechanisms of surfactants at the air–water interface." Advances in Colloid and Interface Science **85**(2–3): 103-144.

Eom, S. H., H. Park, et al. (2010). "High efficiency polymer solar cells via sequential inkjet-printing of PEDOT: PSS and P3HT: PCBM inks with additives." Organic Electronics **11**(9): 1516-1522.

Europe, S. S. E. I. f. (2012). "Alarming situation for european solar companies." from http://www.prosun.org/en/fair-competition/alarming-situation.html.

Florschuetz, L., D. Metzger, et al. (1981). "Streamwise flow and heat transfer distributions for jet array impingement with crossflow."

Galagan, Y., E. W. Coenen, et al. (2012). "Evaluation of ink-jet printed current collecting grids and busbars for ITO-free organic solar cells." Solar Energy Materials and Solar Cells **104**: 32-38.

Galagan, Y., I. G. de Vries, et al. (2011). "Technology development for roll-to-roll production of organic photovoltaics." Chemical Engineering and Processing **50**(5-6): 454-461.

Gonzalez, B., N. Calvar, et al. (2007). "Density, dynamic viscosity, and derived properties of binary mixtures of methanol or ethanol with water, ethyl acetate, and methyl acetate at T=(293.15, 298.15, and 303.15) K." Journal of Chemical Thermodynamics **39**(12): 1578-1588.

Griese, D. (2011). Untersuchung der Beschichtung funktionaler Schichten für organische Leuchtdioden. Institut für Thermische Verfahrenstechnik. Karlsruhe, Karlsruher Institut für Technologie. **Studienarbeit**.

Gutoff, E. B. and E. D. Cohen (2006). Coating and drying defects: troubleshooting operating problems, Wiley-Interscience.

Helgesen, M., R. Sondergaard, et al. (2010). "Advanced materials and processes for polymer solar cell devices." Journal of Materials Chemistry 20(1): 36-60.

Hlawacek, G. (2005). "Kontaktwinkelmessung." Retrieved 24.07., 2013, from http://institute.unileoben.ac.at/physik/Me%DFtechnik/Kontaktwinkelmessung.pdf.

Hoth, C. N., S. A. Choulis, et al. (2009a). "On the effect of poly(3-hexylthiophene) regioregularity on inkjet printed organic solar cells." Journal of Materials Chemistry 19(30): 5398-5404.

Hoth, C. N., R. Steim, et al. (2009b). "Topographical and morphological aspects of spray coated organic photovoltaics." Organic Electronics 10(4): 587-593.

Hotz, C. (2012). "Inbetriebnahme und Einstellung der Prozessparameter einer Batch-Anlage zur inerten Beschichtung von organische Leuchtdioden." Studienarbeit, Karlsuher Institut für Technologie.

Hou, J., M.-H. Park, et al. (2008). "Bandgap and molecular energy level control of conjugated polymer photovoltaic materials based on benzo [1, 2-b: 4, 5-b'] dithiophene." Macromolecules 41(16): 6012-6018.

Hua, X. Y. and M. J. Rosen (1988). "Dynamic surface tension of aqueous surfactant solutions: I. Basic paremeters." Journal of Colloid and Interface Science 124(2): 652-659.

Huang, J., P. F. Miller, et al. (2005). "Investigation of the Effects of Doping and Post-Deposition Treatments on the Conductivity, Morphology, and Work Function of Poly (3, 4-ethylenedioxythiophene)/Poly (styrene sulfonate) Films." Advanced Functional Materials 15(2): 290-296.

Irwin, M. D., D. B. Buchholz, et al. (2008). "p-Type semiconducting nickel oxide as an efficiency-enhancing anode interfacial layer in polymer bulk-heterojunction solar cells." Proceedings of the National Academy of Sciences 105(8): 2783-2787.

Jakubka, F., M. Heyder, et al. (2013). "Determining the coating speed limitations for organic photovoltaic inks." Solar Energy Materials and Solar Cells 109: 120-125.

Jasper, J. J. (1972). "The Surface Tension of Pure Liquid Compounds." Journal of Physical and Chemical Reference Data 1(4): 841-1010.

Jørgensen, M., K. Norrman, et al. (2008). "Stability/degradation of polymer solar cells." Solar Energy Materials and Solar Cells 92(7): 686-714.

Joskow, P. L. and J. E. Parsons (2012). "The future of nuclear power after Fukushima."

Kapur, N. (2003). "A parametric study of direct gravure coating." Chemical Engineering Science 58(13): 2875-2882.

Keddie, J. and A. F. Routh (2010). Fundamentals of latex film formation: processes and properties, Springer.

Khan, A., M. Kasagi, et al. (1982.). Heat transfer augmentation in an axisymmetric impinging jet. Proceedings of the seventh international heat transfer conference., Hemisphere, New York.

Kim, H., W. W. So, et al. (2007a). "The importance of post-annealing process in the device performance of poly(3-hexylthiophene): Methanofullerene polymer solar cell." Solar Energy Materials and Solar Cells 91(7): 581-587.

Kim, S. S., S. I. Na, et al. (2007b). "Efficient polymer solar cells fabricated by simple brush painting." Advanced Materials 19(24): 4410-4415.

Kim, Y. H., C. Sachse, et al. (2011). "Highly Conductive PEDOT: PSS Electrode with Optimized Solvent and Thermal Post-Treatment for ITO-Free Organic Solar Cells." Advanced Functional Materials 21(6): 1076-1081.

Kistler, S. F. and P. M. Schweizer (1997). Liquid Film Coating. London, Chapman and Hall

Kopola, P., T. Aernouts, et al. (2010). "High efficient plastic solar cells fabricated with a high-throughput gravure printing method." Solar Energy Materials and Solar Cells 94(10): 1673-1680.

Kopola, P., M. Tuomikoski, et al. (2009). "Gravure printed organic light emitting diodes for lighting applications." Thin Solid Films 517(19): 5757-5762.

Krebs, F. C. (2009a). "Fabrication and processing of polymer solar cells: A review of printing and coating techniques." Solar Energy Materials and Solar Cells 93(4): 394-412.

Krebs, F. C. (2009b). "Polymer solar cell modules prepared using roll-to-roll methods: Knife-over-edge coating, slot-die coating and screen printing." Solar Energy Materials and Solar Cells 93(4): 465-475.

Krebs, F. C. (2009c). "Roll-to-roll fabrication of monolithic large-area polymer solar cells free from indium-tin-oxide." Solar Energy Materials and Solar Cells 93(9): 1636-1641.

Krebs, F. C., M. Jorgensen, et al. (2009). "A complete process for production of flexible large area polymer solar cells entirely using screen printing-First public demonstration." Solar Energy Materials and Solar Cells 93(4): 422-441.

Kumar, S. and G. D. Scholes (2008). "Colloidal nanocrystal solar cells." Microchimica Acta 160(3): 315-325.

Kumberg, J. (2012). Thermische Behandlung von Polymer-Nanopartikel-Filmen zur Verbesserung der opto-elektrischen Eigenschaften in Polymersolarzellen. Institut für Thermische Verfahrenstechnik. Karlsruhe, Universität Karlsruhe - KIT (TH). **Bachelorarbeit**.

Lee, K.-Y., L.-D. Liu, et al. (1992). "Minimum wet thickness in extrusion slot coating." Chemical Engineering Science 47(7): 1703-1713.

Li, G., V. Shrotriya, et al. (2005). "High-efficiency solution processable polymer photovoltaic cells by self-organization of polymer blends." Nat Mater 4(11): 864-868.

Lightfood, T. (2009). Gravure Coating. European Coating Symposium : Short Course. Karlsruhe, Germany.

Lin, Y., Y. Li, et al. (2012). "Small molecule semiconductors for high-efficiency organic photovoltaics." Chemical Society Reviews 41(11): 4245-4272.

Lunt, R. R., N. C. Giebink, et al. (2009). "Exciton diffusion lengths of organic semiconductor thin films measured by spectrally resolved photoluminescence quenching." Journal of Applied Physics 105(5): 053711-053711-053717.

Ma, C. (2012). Charakterisierung von Elektroden für Lithium-Ionen-Batterien - Parameterstudie, Karlsruher Institut für Technologie. **Diplomarbeit**.

Ma, W., C. Yang, et al. (2005). "Thermally stable, efficient polymer solar cells with nanoscale control of the interpenetrating network morphology." Advanced Functional Materials 15(10): 1617-1622.

Machui, F., S. Langner, et al. (2012). "Determination of the P3HT: PCBM solubility parameters via a binary solvent gradient method: Impact of solubility on the photovoltaic performance." Solar Energy Materials and Solar Cells 100: 138-146.

Macosko, C. and R. Larson (1994). Rheology: principles, measurements, and applications. New York, WILEY-VCH.

Martin, H. (2002). Gk1-6: Wärmeübertragung in Prallströmung. VDI-Wärmeatlas: Berechnungsunterlagen Für Druckverlust, Wärme-und Stoffübergang. Berlin, Springer DE. 1.

Mihailetchi, V. D., H. X. Xie, et al. (2006). "Origin of the enhanced performance in poly(3-hexylthiophene): [6,6]-phenyl C-61-butyric acid methyl ester solar cells upon slow drying of the active layer." Applied Physics Letters 89(1): -.

Miller, S., G. Fanchini, et al. (2008). "Investigation of nanoscale morphological changes in organic photovoltaics during solvent vapor annealing." Journal of Materials Chemistry 18(3): 306-312.

National-Geographic-Online. (2012). "Solar Energy." Retrieved 19.11., from http://environment.nationalgeographic.com/environment/global-warming/solar-power-profile/.

Owens, D. K. and R. C. Wendt (1969). "Estimation of Surface Free Energy of Polymers." Journal of Applied Polymer Science 13(8): 1741-&.

Park, H. J., M. G. Kang, et al. (2010). "A Facile Route to Polymer Solar Cells with Optimum Morphology Readily Applicable to a Roll-to-Roll Process without Sacrificing High Device Performances." Advanced Materials.

Pearson, A. J., T. Wang, et al. (2012). "Rationalizing phase transitions with thermal annealing temperatures for P3HT: PCBM organic photovoltaic devices." Macromolecules 45(3): 1499-1508.

Peet, J., M. L. Senatore, et al. (2009). "The Role of Processing in the Fabrication and Optimization of Plastic Solar Cells." Advanced Materials 21(14-15): 1521-1527.

Peters, K. (2013). Liquid film coating of small molecule OLEDs. Institut für Thermische Verfahrenstechnik - Thin Film Technology. Karlsruhe, Karlsruher Institut für Technologie (KIT). **Dr.-Ing.**

Roesch, R., K.-R. Eberhardt, et al. (2013). "Polymer solar cells with enhanced lifetime by improved electrode stability and sealing." Solar Energy Materials and Solar Cells 117: 59-66.

Romero, O., W. Suszynski, et al. (2004). "Low-flow limit in slot coating of dilute solutions of high molecular weight polymer." Journal of Non-Newtonian Fluid Mechanics 118(2): 137-156.

Ruschak, K. J. (1976). "Limiting flow in a pre-metered coating device." Chemical Engineering Science 31(11): 1057-1060.

Sanyal, M., B. Schmidt-Hansberg, et al. (2011a). "Effect of photovoltaic polymer/fullerene blend composition ratio on microstructure evolution during film solidification investigated in real time by X-ray diffraction." Macromolecules 44(10): 3795-3800.

Sanyal, M., B. Schmidt-Hansberg, et al. (2011b). "In Situ X-Ray Study of Drying-Temperature Influence on the Structural Evolution of Bulk-Heterojunction Polymer–Fullerene Solar Cells Processed by Doctor-Blading." Advanced Energy Materials 1(3): 363-367.

Schabel, W. and H. Martin (2010). G10 Impinging Jet Flow Heat Transfer. VDI Heat Atlas, Springer: 745-752.

Scharber, M. C., D. Mühlbacher, et al. (2006). "Design rules for donors in bulk-heterojunction solar cells—Towards 10% energy-conversion efficiency." Advanced Materials 18(6): 789-794.

10. References

Schlegel, U. (2011). Numerische Untersuchungen des Stoffübergangs in Prallstrahlsystemen bei simultaner Absaugung. Institut für Thermische Verfahrenstechnik, Karlsruher Institut für Technologie.

Schmidt-Hansberg, B. (2012). Process-structure-property Relationship of Polymer-fullerene Bulk Heterojunction Films for Organic Solar Cells: Drying Process, Film Structure and Optoelectronic Properties, Cuvillier.

Schmidt-Hansberg, B., M. Baunach, et al. (2011a). "Spatially resolved drying kinetics of multi-component solution cast films for organic electronics." Chemical Engineering and Processing: Process Intensification 50(5): 509-515.

Schmidt-Hansberg, B., H. Do, et al. (2009a). "Drying of thin film polymer solar cells." European Physical Journal-Special Topics 166: 49-53.

Schmidt-Hansberg, B., M. F. G. Klein, et al. (2009b). "In situ monitoring the drying kinetics of knife coated polymer-fullerene films for organic solar cells." Journal of Applied Physics 106(12): -.

Schmidt-Hansberg, B., M. Sanyal, et al. (2012). "Investigation of non-halogenated solvent mixtures for high throughput fabrication of polymer–fullerene solar cells." Solar Energy Materials and Solar Cells 96: 195-201.

Schmidt-Hansberg, B., M. Sanyal, et al. (2011b). "Moving through the phase diagram: morphology formation in solution cast polymer–fullerene blend films for organic solar cells." Acs Nano 5(11): 8579-8590.

Schmidt-Hansberg, B., M. Sanyal, et al. (2011c). "Moving through the Phase Diagram: Morphology Formation in Solution Cast Polymer–Fullerene Blend Films for Organic Solar Cells." Acs Nano 5(11): 8579-8590.

Schmitt, M., M. Baunach, et al. (2013). "Slot-die processing of lithium-ion battery electrodes— Coating window characterization." Chemical Engineering and Processing: Process Intensification 68(0): 32-37.

Schrödner, M., S. Sensfuss, et al. (2012). "Reel-to-reel wet coating by variation of solvents and compounds of photoactive inks for polymer solar cell production." Solar Energy Materials and Solar Cells 107(0): 283-291.

Service, R. F. (2011). "Outlook Brightens for Plastic Solar Cells." Science 332(6027): 293.

Shaheen, S. E., C. J. Brabec, et al. (2001a). "2.5% efficient organic plastic solar cells." Applied Physics Letters 78(6): 841-843.

Shaheen, S. E., R. Radspinner, et al. (2001b). "Fabrication of bulk heterojunction plastic solar cells by screen printing." Applied Physics Letters 79(18): 2996-2998.

Sharma, A. and E. Ruckenstein (1990). "Energetic criteria for the breakup of liquid films on nonwetting solid surfaces." Journal of Colloid and Interface Science 137(2): 433-445.

Sigma-Aldrich. (2013). "Product Category: Organic and Printed Electronics." Retrieved 12.07., 2013, from http://www.sigmaaldrich.com.

Smits, F. (1958). "Measurement of sheet resistivities with the four-point probe." Bell Syst. Tech. J 37(3): 711-718.

Snoeijer, J. H. and B. Andreotti (2013). "Moving contact lines: Scales, regimes, and dynamical transitions." Annual review of fluid mechanics 45: 269-292.

Solarwirtschaft, B. (2013). "Photovoltaik Preisindex." Retrieved 01.06., 2013, from http://www.solarwirtschaft.de/preisindex.

Søndergaard, R., M. Hösel, et al. (2012a). "Roll-to-roll fabrication of polymer solar cells." Materials Today 15(1): 36-49.

Søndergaard, R., M. Hösel, et al. (2012b). "Roll-to-roll fabrication of polymer solar cells." Materials Today 15(1–2): 36-49.

Søndergaard, R. R., M. Hösel, et al. (2013). "Roll-to-Roll fabrication of large area functional organic materials." Journal of Polymer Science Part B: Polymer Physics 51(1): 16-34.

Stephan, I. K. (1994). Konvektiver Wärme-und Stoffübergang. Strömungen mit Phasenumwandlungen. Wärme-und Stoffübertragung, Springer: 415-515.

Sun, B. and H. Sirringhaus (2006). "Surface tension and fluid flow driven self-assembly of ordered ZnO nanorod films for high-performance field effect transistors." Journal of the American Chemical Society 128(50): 16231-16237.

Susanna, G., L. Salamandra, et al. (2011). "Airbrush spray-coating of polymer bulk-heterojunction solar cells." Solar Energy Materials and Solar Cells 95(7): 1775-1778.

Takano, T., H. Masunaga, et al. (2012). "PEDOT Nanocrystal in Highly Conductive PEDOT: PSS Polymer Films." Macromolecules 45(9): 3859-3865.

Teja, A. S. and P. Rice (1981). "Generalized Corresponding States Method for the Viscosities of Liquid-Mixtures." Industrial & Engineering Chemistry Fundamentals 20(1): 77-81.

Thon, A. H., S. M., et al. (2012). "Hybrid passivated colloidal quantum dot solids." Nat Nano 7(9): 577-582.

Vak, D. J., S. S. Kim, et al. (2007). "Fabrication of organic bulk heterojunction solar cells by a spray deposition method for low-cost power generation." Applied Physics Letters 91(8): -.

Vattenfall-AB. (2013). "Moabit Fakten." Retrieved 24.06., 2013, from http://kraftwerke.vattenfall.de/powerplant/moabit.

VDI (2002). VDI-Wärmeatlas: Berechnungsblätter für den Wärmeübergang. Düsseldorf, Vdi-Verlag.

Vogt, B. D., C. L. Soles, et al. (2005). "Moisture absorption into ultrathin hydrophilic polymer films on different substrate surfaces." Polymer 46(5): 1635-1642.

Voigt, M. M., R. C. I. Mackenzie, et al. (2012). "Gravure printing inverted organic solar cells: The influence of ink properties on film quality and device performance." Solar Energy Materials and Solar Cells 105: 77-85.

Walker, S. B. and J. A. Lewis (2012). "Reactive silver inks for patterning high-conductivity features at mild temperatures." Journal of the American Chemical Society 134(3): 1419-1421.

Walker, W. H., W. K. Lewis, et al. (1923). "Principles of chemical engineering."

Wang, J., X. Ren, et al. (2011). "Charge accumulation induced S-shape< i> J</i>–< i> V</i> curves in bilayer heterojunction organic solar cells." Organic Electronics 12(6): 880-885.

Wengeler, L., K. Peters, et al. (2013). "Fluid-dynamic properties and wetting behavior of coating inks for roll-to-roll production of polymer-based solar cells." Journal of Coatings Technology and Research: 1-9.

Wengeler, L. and W. Schabel (2010). "Verfahrenstechnische Herausforderungen für die Herstellung großflächiger organischer und hybrider Solarzellen." Nanotechnik/ Photonik, AT-Fachverlag, Stuttgart(August 2010, No-4).

Wengeler, L., B. Schmidt-Hansberg, et al. (2011). "Investigations on knife and slot die coating and processing of polymer nanoparticle films for hybrid polymer solar cells." Chemical Engineering and Processing: Process Intensification 50(5-6): 478-482.

10. References

Wengeler, L., M. Schmitt, et al. (2012). "Comparison of large scale coating techniques for organic and hybrid films in polymer based solar cells." Chemical Engineering and Processing: Process Intensification(0).

Wenz, T. (2012). Benetzungseigenschaften von Flüssigkeiten und Oberflächen für polymerbasierte Solarzellen. Institut für Thermische Verfahrenstechnik, Universität Karlsruhe - KIT (TH). **Studienarbeit**.

Willenbacher, N. (2013). Rheologie von Beschichtungsfluiden. Hochschulkurs: Beschichtung und Trocknung von dünnen Schichten. Karlsruhe, Germany.

Willinger, B. (2013). Theoretische und fluiddynamische Betrachtung von Beschichtungsströmungen. Hochschulkurs Beschichten und Trocknen von dünnen Schichten. Karlsruhe, Germany.

Wilson, G. (2013). "Research cell efficiency records." Retrieved 25.05., 2013, from http://www.nrel.gov/ncpv/images/efficiency_chart.jpg.

Wolf, R. and A. C. Sparavigna (2010). "Role of Plasma Surface Treatments on Wetting and Adhesion." Engineering 2(6): 397-402.

Wu, Y. and G. Zhang (2010). "Performance enhancement of hybrid solar cells through chemical vapor annealing." Nano Letters 10(5): 1628-1631.

Xia, Y. J., K. Sun, et al. (2012). "Solution-Processed Metallic Conducting Polymer Films as Transparent Electrode of Optoelectronic Devices." Advanced Materials 24(18): 2436-2440.

Yang, H., T. J. Shin, et al. (2005). "Effect of mesoscale crystalline structure on the field-effect mobility of regioregular poly (3-hexyl thiophene) in thin-film transistors." Advanced Functional Materials 15(4): 671-676.

Yaws, C. (1999). Chemical properties handbook: physical thermodynamic, environmental, transport, safety and health related properties for organic and inorganic chemicals. New York, McGraw-Hill.

Zhou, Y. F., F. S. Riehle, et al. (2010). "Improved efficiency of hybrid solar cells based on non-ligand-exchanged CdSe quantum dots and poly(3-hexylthiophene)." Applied Physics Letters 96(1): -.

Zimmermann, B., H. F. Schleiermacher, et al. (2011). "ITO-free flexible inverted organic solar cell modules with high fill factor prepared by slot die coating." Solar Energy Materials and Solar Cells 95(7): 1587-1589.